Probabilistic Combinatorial Optimization on Graphs

Probabilistic Combinatorial Optimization on Graphs

Cécile Murat
and
Vangelis Th. Paschos

ISTE

First published in Great Britain and the United States in 2006 by ISTE Ltd

ISTE Ltd
6 Fitzroy Square
London W1T 5DX
UK
www.iste.co.uk

ISTE USA
4308 Patrice Road
Newport Beach, CA 92663
USA

First South Asian Edition 2007

Library of Congress Cataloging-in-Publication Data

Murat, Cecile.
Probabilistic combinatorial optimization on graphs / Cecile Murat and Vangelis Th. Paschos.
p. cm.
Includes bibliographical references and index.
ISBN 13: 978-1-905209-33-0
ISBN 10: 1-905209-33-9
1. Combinatorial probabilities. 2. Combinatorial optimization. 3. Random graphs. I. Paschos, Vangelis Th. II. Title.
QA273.45.M87 2006
519.2—dc22
2006000868

British Library Cataloguing-in-Publication Data
A CIP record for this book is available from the British Library
ISBN 10: 1-905209-33-9
ISBN 13: 978-1-905209-33-0

Printed in Brijbasi Art Press Ltd., I-72, Sector-9, Noida, U.P. India.

Contents

Preface

This monograph is the outcome of our work on probabilistic combinatorial optimization since 1994. The first time we heard about it, it seemed to us to be a quite strange scientific area, mainly because randomness in graphs is traditionally expressed by considering probabilities on the edges rather than on the vertices. This strangeness was our first motivation to deal with probabilistic combinatorial optimization. As our study progressed, we have discovered nice mathematical problems, connections with other domains of combinatorial optimization and of theoretical computer science, as well as powerful ways to model real-world situations in terms of graphs, by representing reality much more faithfully than if we do not use probabilities on the basic data describing them, i.e., the vertices.

What is probabilistic combinatorial optimization? Basically, it is a way to deal with aspects of robustness in combinatorial optimization. The basic problematic is the following. We are given a graph (let us denote it by $G(V, E)$, where V is the set of its points, called vertices, and E is a set of straight lines, called edges, linking some pairs of vertices in V), on which we have to solve some optimization problem Π. But, for some reasons depending on the reality modelled by G, Π is only going to be solved for some subgraph G' of G (determined by the vertices that will finally be present) rather than for the whole of G. The measure of how likely it is that a vertex $v_i \in V$ will belong to G' (i.e., will be present for the final optimization) is expressed by a probability p_i associated with v_i. How we can proceed in order to solve Π under this kind of uncertainty?

A first very natural idea that comes to mind is that one waits until G' is specified (i.e., it is present and ready for optimization) and, at this time, one solves Π in G'. This is what is called *re-optimization*. But what if there remains very little time for such a computation? We arrive here at the basic problematic of the book. If there is no time for re-optimization, another way to proceed is the following. One solves Π in the whole of G in order to get a feasible solution (denoted by S), called *a priori solution*, which will serve her/him as a kind of benchmark for the solution on the effectively present subgraph G'. One has also to be provided with an algorithm that modifies S in order to fit G'. This algorithm is called *modification strategy* (let us denote it by `M`). The objective now becomes to compute an *a priori* solution that, when modified by `M`, remains "good" for any subgraph of G (if this subgraph is the one where Π will be finally solved). This amounts to computing a solution that optimizes a kind of expectation of the size of the modification of S over all the possible subgraphs of G, i.e., the sum of the products of the probability that G' is the finally present graph multiplied by the value of the modification of S in order to fit G' over any subgraph G' of G. This expectation, depending on both the instance of the deterministic problem Π, the vertex-probabilities, and the modification strategy adopted, will be called the *functional*. Obviously, the presence-probability of G' is the probability that all of its vertices are present.

Seen in this way, the probabilistic version PΠ of a (deterministic) combinatorial optimization problem Π becomes another equally deterministic problem Π', the solutions of which have the same feasibility constraints as those of Π but with a different objective function where vertex-probabilities intervene. In this sense, probabilistic combinatorial optimization is very close to what in the last couple of years has been called "one stage optimisation under independent decision models", an area very popular in the stochastic optimization community.

What are the main mathematical problems dealing with probabilistic consideration of a problem Π in the sense discussed above? We can identify at least five interesting mathematical and computational problems dealing with probabilistic combinatorial optimization:

1) write the functional down in an analytical closed form;

2) if such an expression of the functional is possible, prove that its value is polynomially computable (this amounts to proving that the modified problem Π' belongs to **NP**);

3) determine the complexity of the computation of the optimal *a priori* solution, i.e., of the solution optimizing the functional (in other words, determine the computational complexity of Π');

4) if Π' is **NP**-hard, study polynomial approximation issues;

5) always, under the hypothesis of the **NP**-hardness of Π', determine its complexity in the special cases where Π is polynomial, and in the case of **NP**-hardness, study approximation issues.

Let us note that, although curious, point 2 in the above list in neither trivial nor senseless. Simply consider that the summation for the functional includes, in a graph of order n, 2^n terms (one for each subgraph of G). So, polynomiality of the computation of the functional is, in general, not immediate.

Dealing with the contents of the book, in Chapter 1 probabilistic combinatorial optimization is formally introduced and some old relative results are quickly presented.

The rest of the book is subdivided into two parts. The first one (Part I) is more computational, while the second (Part II) is rather "structural". In Part I, after formally introducing probabilistic combinatorial optimization and presenting some older results (Chapter 1), we deal with probabilistic versions of four paradigmatic combinatorial problems, namely, PROBABILISTIC MAX INDEPENDENT SET, PROBABILISTIC MIN VERTEX COVER, PROBABILISTIC LONGEST PATH and PROBABILISTIC MIN COLORING (Chapters 2, 3, 4 and 5, respectively). For any of them, we try, more or less, to solve the five types of problems just mentioned.

As the reader will see in what follows, even if, mainly in Chapters 2 and 3, several modification strategies are used and analyzed, the strategy that comes back for all the problems covered is the one consisting of moving absent vertices out of the *a priori* solution (it is denoted by MS for the rest of the book). Such a strategy is very quick, simple and intuitive but it does not always produce feasible solutions for any of the possible subgraphs (i.e., it is not always feasible). For instance, if it is feasible for PROBABILISTIC MAX INDEPENDENT SET, PROBABILISTIC MIN VERTEX COVER and PROBABILISTIC MIN COLORING, this is not the case for PROBABILISTIC LONGEST PATH, unless particular structure is assumed for the input graph. So, in Part II, we restrict ourselves to this particular strategy and assume that either MS is feasible, or, in case of unfeasibility, very little additional work is required in order to achieve feasible solutions. Then, for large classes of problems (e.g., problems where feasible solutions are subsets of the initial vertex-set or edge-set satisfying particular properties, such as stability, etc.), we investigate relations between these problems and their probabilistic counterparts (under MS). Such relations very frequently derive answers to the above mentioned five types of problems. Chapter 7 goes along the same lines as Chapter 6. We present a small compendium of probabilistic graph-problems (under MS). More precisely we revisit the most well-known and well-studied graph-problems and we investigate if strategy MS is feasible for any of them. For the problems for which this statement holds, we express the functional associated with it and, when possible, we characterize the optimal *a priori* solution and the complexity of its computation.

The book should be considered to be a monograph as in general it presents the work of its authors on probabilistic combinatorial optimization graph-problems. Nevertheless, we think that when the interested readers finish reading, they will be perfectly aware of the principles and the main issues of the whole subject area. Moreover, the book aims at being a self-contained work, requiring only some mathematical maturity and some knowledge about complexity and approximation theoretic notions. For help, some appendices have been added, dealing, on the one hand, with some mathematical preliminaries: on sets, relations and functions, on basic concepts from graph-theory and on some elements from discrete probabilities and, on the other hand, with elements of the complexity and the polynomial approximation theory: notorious complexity classes, reductions and **NP**-completeness and basics about the polynomial approximation of **NP**-hard problems. We hope that with all that, the reader will be able to read the book without much preliminary effort. Let us finally note that, for simplifying reading of the book, technical proofs are placed at the end of each chapter.

As we have mentioned in the beginning of this preface, we have worked in this domain since 1994. During all these years many colleagues have read, commented, improved and contributed to the topics of the book. In particular, we wish to thank Bruno Escoffier, Federico Della Croce and Christophe Picouleau for having working with, and encouraged us to write this book. The second author warmly thanks Elias Koutsoupias and Vassilis Zissimopoulos for frequent invitations to the University of Athens, allowing full-time work on the book, and for very fruitful discussions. Many thanks to Stratos Paschos for valuable help on LaTeX.

Tender and grateful thanks to our families for generous and plentiful support and encouragement during the task.

Finally, it is always a pleasure to work with Chantal and Sami Menasce, Jon Lloyd and their colleagues at ISTE.

Cécile Murat and Vangelis Th. Paschos
Athens and Paris, October 2005

Chapter 1

A Short Insight into Probabilistic Combinatorial Optimization

1.1. Motivations and applications

The most common way in which probabilities are associated with combinatorial optimization problems is to consider that the data of the problem are deterministic (always present) and randomness carries over the relation between these data (for example, randomness on the existence of an edge linking two vertices in the framework of a random graph theory problem ([BOL 85]) or randomness on the fact that an element is included to a set or not, when dealing with optimization problems on set-systems or, even, randomness on the execution time of a task in scheduling problems). Then, in order to solve an optimization problem, algorithms (probabilistic or, more frequently, deterministic) are devised, and the mathematical expectation of the obtained solution is measured. A main characteristic of this approach is that probabilities do not intervene in the mathematical formulation of the problems but only in the mathematical analysis performed in order to get results.

More recently, in the late 1980s, another approach to the randomness of combinatorial optimization problems was developed: probabilities are associated with the data describing an optimization problem (for a particular datum, we can see the probability associated with it as a measure of how much this datum is likely to be present in the instance to be finally optimized) and, in this sense, probabilistic elements are explicitly included in the formulations of these problems. Such formulations give rise to what we will call *probabilistic combinatorial optimization problems*. Here, the objective function is a form of carefully defined mathematical expectation over all possible instances of size less than, or equal to, a given initial size.

The fact that, when dealing with probabilistic combinatorial problems, randomness lies in the presence of the data means that the underlying models are very suitable for the modelling of natural problems, where randomness is the quantification of uncertainty, or fuzzy information, or inability to forecast phenomena, etc.

For instance, in several versions of satellite shot planning problems, the uncertainty concerning meteorological conditions can be quantified by a system of probabilities. The optimization problems derived are, as we will see later in this chapter, clearly of probabilistic nature. If, on the other hand, during a salesman's tour, some clients need not to be visited, he should omit them from his tour and if the fact that a client has to be visited or not is modelled in terms of probabilities-systems, then a probabilistic traveling salesman problem arises[1]. For similar or other reasons, starting from a transportation, or computer, or any other kind of network, we encounter problems like probabilistic shortest path problem[2] or probabilistic longest path problem[3], probabilistic minimum spanning tree problem[4], etc. Also, in industrial automation, the systems for foreseeing workshops' production give rise to probabilistic scheduling, or probabilistic set covering or probabilistic set packing, etc. Finally, in computer science, mainly when dealing with parallelism or distributed computation, probabilistic combinatorial optimization problems very frequently have to be solved. For instance, modeling load-balancing with non-uniform processors and failures possibility

1. Given a complete graph on n vertices, denoted by K_n, with positive distances on its edges, the minimum traveling salesman problem consists of minimizing the cost of a Hamiltonian cycle (see section A.2 in Appendix A), the cost of such a cycle being the sum of the distances of its edges.

2. Given an edge-weighted directed or undirected graph $G(V, E, \vec{w})$ and either a fixed vertex s (first version), or two fixed vertices s and t (second version), the objective for the first version is to determine minimum-weight paths from s to any other vertex of G; for the second version, the objective is to determine a minimum-weight path from s to t, the weight of a path being the sum of the weights of its edges; the most famous variants of these versions are the ones where G is assumed directed; in this case, the three most-known polynomial algorithms solving the first version of the problem are the ones of Dijkstra ([DIJ 59]), under the assumption that edge-weights are non-negative, and of Bellman ([BEL 57]), under the assumption that the input-graph contains no circuit; no algorithm is known for the second version of the problem that is solved as a special case of the first version; a third version of the shortest path problem, where we search for shortest paths for any pair of vertices, is polynomially solved by the algorithm of Floyd ([FLO 62]).

3. Given a graph $G(V, E)$, an edge-weight function $w : E \rightarrow \mathbb{Q}$, and two specific vertices s and t, the longest path problem consists of determining a maximum total-weight simple path from s to t, the total weight of a path being the sum of the weights of its edges.

4. Given an edge-weighted undirected graph $G(V, E, \vec{w})$, the objective is to determine a minimum-weight tree spanning V, the weight of such a tree being the sum of the weights of its edges; the most famous polynomial algorithm solving this problem is the one of Kruskal ([KRU 56]).

becomes a probabilistic graph partitioning problem; also in network reliability theory, many probabilistic routing problems are met ([BER 90b]).

In all, models of probabilistic nature are very suitable and appropriate real-life problems where randomness is a constant source of concern and, on the other hand, the study of the problems derived from these models are very attractive as mathematical abstraction of real systems. Another reason motivating work on probabilistic combinatorial optimization is the study and the analysis of the stability of the optimal solutions of deterministic combinatorial optimization problems when the considered instances are perturbed. For problems defined on graphs, more particularly, these perturbations are simulated by the occurrence, or the absence, of subsets of vertices (see, for example, [PEE 99] where probabilistic combinatorial optimization approaches and concepts are used to yield robust solutions for an on-line traffic-assignment problem).

Informally, given a combinatorial optimization graph[5]-problem Π, defined on a graph $G(V, E)$, an instance of its probabilistic counterpart, denoted by PΠ, is built by associating a probability p_i with any vertex v_i in V. This probability is considered as the presence-probability of v_i in the subgraph of G on which Π will be finally solved. Problem PΠ expresses the fact that Π will, eventually, have to be solved not on the whole G, but rather on some of its subgraphs that will be specified very shortly before the solution in this subgraph is required.

In order to illustrate the issue outlined above, we will consider in what follows four examples of models that give rise to probabilistic combinatorial optimization problems.

EXAMPLE 1.1.– *Probabilistic traveling salesman.* A repair company has to perform a minimum-length daily tour visiting n potential clients. This is the classical (deterministic) traveling salesman problem, denoted by MIN TSP in what follows. It is formally defined as follows: given a set C of n cities and distances $d(c_i, c_j) \in \mathbb{Q}$, for any pair $(c_i, c_j) \in C \times C$, $i \neq j$, MIN TSP consists of determining a tour of C, i.e., a permutation $\sigma : \{1, \ldots, n\} \rightarrow \{1, \ldots, n\}$, minimizing the length of the tour, i.e., the quantity $d(c_{\sigma(n)}, c_{\sigma(1)}) + \sum_{i=1}^{n-1} d(c_{\sigma(i)}, c_{\sigma(i+1)})$. MIN TSP is commonly modeled in terms of a complete graph, denoted by K_n (see section A.2 of Appendix A) on n vertices (representing the cities). Edge (v_i, v_j) is weighted by $d(c_i, c_j)$ and an optimal solution is a Hamiltonian cycle (section A.2 of Appendix A) of minimum total length (or distance), the length of a cycle being the sum of the distances on its edges. But, if we assume that any client will not need to be repaired every day, then this implies that, a given date, only a subset of clients need to be effectively visited; this subset changes from day to day. What can be done is that a client i can be assigned, for a

5. In this book, only graph-problems are considered; for probabilistic combinatorial optimization problems defined on other data structures, the interested reader can be referred, for example, to [BEL 93] where some scheduling problems, as well as the bin-packing problem, are studied.

random day, with a repairing-probability p_i; this probability is independent from the probabilities dealing with the other clients. We thus get a version of the probabilistic traveling salesman problem (initially introduced and studied in [JAI 85, JAI 88a]).

EXAMPLE 1.2.– *Probabilistic coloring.* Consider for a given University-fall a list of potential classes that students can follow: any student has to choose a sublist of such classes. For any of them, one knows the title, the teaching professor and the time slot assigned to it, each such slot being proposed by the professor in charge. A class will only take place if the number of students having chosen it is above a given threshold. So, nobody knows *a priori* if a particular class will take place before the closing of students' registrations (we can reasonably assume that the choice of any student is dependent on the contents of the course and of the teacher). On the other hand, one can, for example, by looking at statistics on the behavior of the students during past years, assign probabilities on the fact that a particular class will really take place, the mandatory courses being assigned with probability 1. The problem for the University planning services is how many rooms need to be assigned to the set of courses. This problem is typically an instance of probabilistic coloring if one considers courses as vertices and if one links two such vertices if the corresponding classes cannot take place in the same room (because they are planned with the same professor, or are assigned with overlapping time slots). This type of graph is known by the term incompatibility graph. Here, an independent set, i.e., a potential color, corresponds to a set of "compatible classes", i.e., to classes that can be assigned with the same room. The number of colors used in such a graph represents the total number of rooms assigned to the set of classes considered. The probabilities resulting from the statistical analysis on the former students' behavior are the presence probabilities for the vertices (i.e., the probabilities that the corresponding classes will really take place).

EXAMPLE 1.3.– *Probabilistic independent set.* Consider a planning aiding process for realizing satellite shots. One associates a vertex with any shot requested and one links two vertices if they correspond to shots that cannot be realized on the same orbit. But a shot realized under, for example, strong cloud cover cannot be used for the purposes for which it has been requested. Using meteorological forecasting, one can assign to any shot requested a probability that it will be usable. This problem has been modelled in [GAB 97] (see also [GAB 94]) as a maximum independent set and if one takes into account probabilities associated with meteorological forecasting, then one has to solve a probabilistic version of it.

EXAMPLE 1.4.– *Probabilistic longest path.* For the satellite shot planning problem dealt in Example 1.3, one can use another graph-representation (see [GAB 97] for details) where an arc models the possibility of successive realization of its two end-points. Then, the satellite shot planning can be represented as a particular longest path problem. Integration of probabilities associated, for instance, with meteorological forecasting to this model gives rise to a probabilistic longest path.

1.2. A formalism for probabilistic combinatorial optimization

We have already mentioned that the probabilistic version of an optimization problem models the fact that, given an instance of the problem, only a subinstance of it will eventually be solved. Since we do not know which is this subinstance, the most natural approach that comes in mind is to optimally solve any particular subinstance of the problem at hand (following the probabilities on its vertices, any such subinstance is more or less likely to be the one where optimization has to be effectively performed). Such an approach, called reoptimization in [BER 90b, BER 93, JAI 93], can be very much time- and space-consuming, in particular when the initial problem is **NP**-hard. Indeed, given a graph $G(V, E)$ of order n (i.e., $|V| = n$), there exist 2^n subsets of V and consequently 2^n subgraphs, induced by these subsets, any of them candidate to be the instance effectively under consideration. For an **NP**-hard problem (this remains true even for a polynomial problem), the amount of time needed to solve any of these instances to the optimum can be huge so that reoptimization becomes practically unrealistic.

This is the main reason for which another, more realistic, approach is used and this will be dealt in this book. It is called an *a priori optimization* and has been introduced in [JAI 85, BER 88]. Informally, instead of reoptimizing any subinstance, the underlying idea of an *a priori* optimization consists of determining a solution of the whole (initial) instance, i.e., the one where all data are present, called an *a priori solution*, and to apply a strategy, called a *modification strategy*, making it possible to adapt as quickly as possible the *a priori* solution to the subinstance that must effectively be solved. The choice of this strategy depends strongly on the application modelled by the problem.

Consider a graph $G(V, E)$ instance of a combinatorial optimization problem Π, a feasible solution S, for Π in G, a subset V' of V and the subgraph $G[V']$ of G induced by V'. A modification strategy `M` is an algorithm transforming S in order to get a feasible Π-solution for any such $G[V']$. Obviously, it is assumed that if `M` is applied on G (i.e., if $V' = V$), then S remains unchanged. Also, one can suppose that application of `M` in the final instance is possible, in the sense that there exists sufficient time for its achievement.

EXAMPLE 1.1. (CONTINUED) – Revisit the repairing company dealt in Example 1.1. Assume that for several material reasons, its staff do not wish to reoptimize the daily tours for its vehicles. A possible way to plan these tours is the following. One computes firstly a feasible tour T including the whole set of the clients is computed; this is the *a priori* solution mentioned just above. A possible modification strategy in order to compute the effective tour for a given day is to drop absent clients (i.e., clients that do not ask for intervention during this day); then, it suffices to visit the present ones following the order induced by the *a priori* tour T.

In practice, the *a priori* approach corresponds to a behavior observed for the real problem; the modification strategy algorithmically models this behavior. The choice of a modification strategy depends strongly on the real-world application modelled. In order to illustrate this, consider the following example inspired from a vehicle routing problem studied in [BER 90b, BER 92, BER 96].

EXAMPLE 1.5.– The problem studied in [BER 90b, BER 92, BER 96] consists of determining a shortest distance tour through n clients under several constraints, for example, limits on vehicle capacities together with assumptions that any of the vehicles have to retrieve different quantities of different objects from any client, etc. If any on day d only a subset of the clients has to be visited, then for modifying an *a priori* tour to fit these present clients, two modification strategies could be used depending on when clients' demands become available:

– In the first strategy, denoted by `M1`, a vehicle, following the *a priori* tour, visits all the clients but it only serves the ones having asked for service on day d. When the vehicle is saturated, i.e., its capacity is attained and it returns to the depot before continuing with the next client.

– The second strategy, denoted by `M2`, differs from `M1` by the fact that the vehicle only visits (following the *a priori* tour) clients having asked for services on day d (returning to the depot when saturated and then continuing with the next client).

In order to illustrate differences between the two strategies, consider an *a priori* tour $(0, 1, 2, 3, 4, 5, 6, 0)$ and assume that depot is vertex 0 and that vehicle has capacity 30. At day d, the clients $1, 4$ and 6 need not to be visited and that the demands for clients 2, 3 and 5 are 20, 10 and 20, respectively. The results for the two strategies above are shown in Figure 1.1.

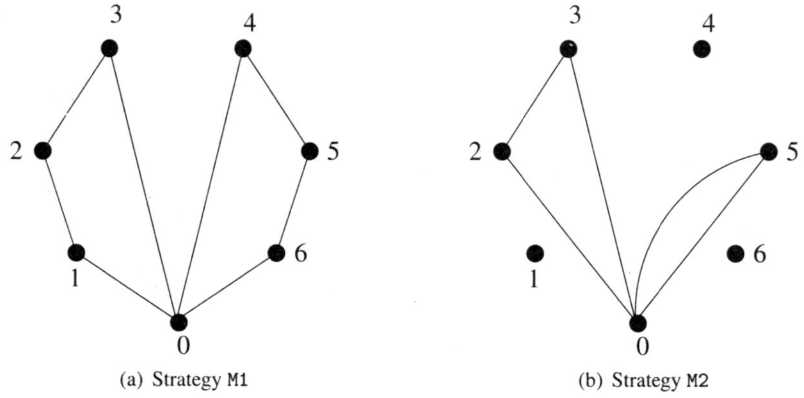

Figure 1.1. *Application of modification strategies `M1` and `M2` for the probabilistic vehicle routing problem with capacity constraints*

As one can see, under strategy M1 (Figure 1.1(a)), the route realized by the vehicle will be $(0, 1, 2, 3, 0)$, then $(0, 4, 5, 6, 0)$, while the route under M2 (Figure 1.1(b)) will be $(0, 2, 3, 0)$, then $(0, 5, 0)$.

There exists an important difference between these two strategies:

– M1 models situations where demand of a particular client becomes clear (or known) only once it has been visited;

– M2 corresponds to situations where clients' demands are known in advance, i.e., before the vehicle starts the route.

A basic operational and computational feature of the *a priori* optimization approach is that the optimization problem considered has to be solved only once; next, the only "tool" needed is a quick modification strategy which is able to adapt the *a priori* solution to the subinstance to be effectively optimized. In this way, computational time is not really a serious problem.

The question now is: "what is the measure of an *a priori* solution?". Let S be a feasible solution for Π on $G(V, E)$, $\mathtt{M}$ be a modification strategy for S and V' be a subset of V. Denote by $S(V', \mathtt{M})$ the solution for Π in $G[V']$, obtained from S by applying $\mathtt{M}$ and by $m(G[V'], S(V', \mathtt{M}))$ its value. A reasonable requirement for $S(V', \mathtt{M})$ is that $m(G[V'], S(V', \mathtt{M}))$ is as close as possible to the value of an optimal solution for Π in $G[V']$, denoted by $\text{opt}(G[V'])$. Since, on the other hand, we do not know *a priori*, which will be the subinstance to be solved, we will use as evaluation-measure for S its expectation. Denote by $\Pr[V']$ the probability of presence of the vertices of V', hence the probability of $G[V']$ and set $\Pr[v_i] = p_i$, the presence-probability of $v_i \in V$; then:

$$\Pr[V'] = \prod_{v_i \in V'} p_i \prod_{v_j \notin V'} (1 - p_j) \qquad [1.1]$$

In particular, when $p_i = p$, for any $v_i \in V$, then [1.1] becomes:

$$\Pr[V'] = p^{|V'|} (1 - p)^{|V| - |V'|}$$

The measure (i.e., the objective function) of S for PΠ, also called *functional* in what follows, is defined as:

$$m(G, S, \mathtt{M}) = E(G, S, \mathtt{M}) = \sum_{V' \subseteq V} m(G[V'], S(V', \mathtt{M})) \Pr[V'] \qquad [1.2]$$

where $\Pr[V']$ is defined by [1.1].

In standard complexity-theoretic language, the problems studied in this book belong to the class **NPO**. Informally, an **NPO** problem is an optimization problem, the decision versions of which is in **NP** (see also Appendix B). More formally now, an **NPO** problem can be defined as follows.

DEFINITION 1.1.– *A problem* Π *in* ***NPO*** *is a quadruple* $(\mathcal{I}_{\Pi}, Sol_{\Pi}, m_{\Pi}, \text{goal}(\Pi))$ *where:*

– $\mathcal{I}_{\Pi}$ *is the set of instances of* Π *(and can be recognized in polynomial time);*

– *given* $I \in \mathcal{I}_{\Pi}$, $Sol_{\Pi}(I)$ *is the set of feasible solutions of* I*; the size of a feasible solution of* I *is polynomial in the size* $|I|$ *of the instance; moreover, one can determine in polynomial time if a solution is feasible or not;*

– *given* $I \in \mathcal{I}_{\Pi}$ *and* $S \in Sol_{\Pi}(I)$, $m_{\Pi}(I, S)$ *denotes the value of the solution* S *of the instance* I*;* m_{Π} *is called the objective function, and is computable in polynomial time;*

– $\text{goal}(\Pi) \in \{\min, \max\}$.

We can now give a formal definition for probabilistic combinatorial optimization problems (under the *a priori* optimization assumption), derived from Definition 1.1.

DEFINITION 1.2.– *Let* $\Pi = (\mathcal{I}_{\Pi}, Sol_{\Pi}, m_{\Pi}, \text{goal}(\Pi))$ *be an* ***NPO*** *problem as in Definition 1.1. The probabilistic version of* Π*, denoted by* $\text{P}\Pi$*, is a six-tuple* $((\mathcal{I}_{\Pi}, \mathbf{Pr}), Sol_{\Pi}, \text{goal}(\Pi), \mathtt{M}, E_{\Pi})$*, where:*

– $\mathcal{I}_{\Pi}$ *is as in Definition 1.1 and* $\mathbf{Pr}$ *is the set of all the vectors* Pr *of the presence-probabilities of the data representing* $I \in \mathcal{I}$*; the pair* $(\mathcal{I}_{\Pi}, \mathbf{Pr})$ *is the instance-set of* $\text{P}\Pi$ *and the couple* $I, \text{Pr}[I]$*,* $I \in \mathcal{I}$*,* $\text{Pr} \in \mathbf{Pr}$ *is an instance of* $\text{P}\Pi$*;* Sol_{Π} *and* $\text{goal}(\Pi)$ *are as in the corresponding items of Definition 1.1;*

– $\mathtt{M}$ *is an algorithm, called modification strategy, such that, given an instance* $(I, \text{Pr}[I])$ *of* Π*, a solution* $S \in Sol(I, \text{Pr}[I])$ *and any subinstance* I' *of* I*, it modifies* S *in order to produce a feasible solution* $S(I', \mathtt{M})$*;*

– E_{Π} *is the functional of* S *and is defined (analogously to [1.2]) as:*

$$E_{\text{P}\Pi}(I, S, \mathtt{M}) = \sum_{I' \subseteq I} m\left(I', S\left(I', \mathtt{M}\right)\right) \text{Pr}\left[I'\right] \qquad [1.3]$$

where $\text{Pr}[I']$ *is defined (analogously to [1.1]) as:*

$$\text{Pr}\left[I'\right] = \prod_{d_i \in I} \text{Pr}\left[d_i\right] \prod_{d_j \notin I} \left(1 - \text{Pr}\left[d_j\right]\right)$$

where d_i*,* d_j *draw data of* I *and* $\text{Pr}[d_i]$ *and* $\text{Pr}[d_j]$ *their presence probabilities respectively.*

One can see that Definition 1.2 implies that modification strategy `M` is part of the definition of the problem. In this sense, two distinct strategies `M1` and `M2`, associated with the same **NPO** problem Π, give rise to two distinct probabilistic problems $\text{P}\Pi_1$ and $\text{P}\Pi_2$, respectively, since changing a modification strategy changes the functional. In other words, distinct modification strategies lead to distinct objective functions.

The modification strategy used most frequently until now is the one consisting of dropping absent data out of the *a priori* solution and of taking the remaining elements of it as a solution for the effective instance. This simple strategy, denoted by MS for the rest of this chapter, is feasible for numerous problems (this is the case of all the problems dealt in this monograph and for the ones dealt in [AVE 94, AVE 95, BEL 93, BER 88, BER 89, BER 90b, JAI 85, JAI 88a, JAI 88b, JAI 92, SÉG 93]) but not for any problem. Let us take for example the case of the probabilistic minimum independent dominating set (also called the minimum maximal independent set). Here, given an *a priori* maximal independent set S, dropping the absent vertices out from S does not necessarily result in a maximal independent set for the present subgraph.

As we will see in the next chapters, in particular under strategy MS and in the cases where the optimum *a priori* solution has a closed combinatorial characterization, the derived probabilistic problems can be equivalently stated as "deterministic combinatorial optimization problems" under particular and sometimes rather non-standard objective functions.

Let us note also that *a priori* optimization under strategy MS corresponds to the following robustness model for combinatorial optimization. Consider a generic instance I of a combinatorial optimization problem Π. Assume that Π is not to be necessarily solved on the whole I, but rather on a (unknown *a priori*) subinstance $I' \subset I$. Suppose that any datum d_i in the data-set describing I has a probability p_i, indicating how d_i is likely to be present in the final subinstance I'. Consider finally that once instance I' is specified, the solver has no opportunity to solve directly instance I'. In this case, there certainly exist many ways to proceed. Here we deal with a simple and natural way where one computes an initial solution S for Π in the entire instance I and, once I' becomes known, one removes from S those elements of S that do not belong to I' (providing that this deletion results in a feasible solution for I') thus giving a solution S' fitting I'. The objective is to determine an initial solution S for I such that, for any subinstance $I' \subseteq I$ presented for optimization, the solution S' respects some predefined quality criterion (for example, optimal for I', or achieving, say, constant approximation ratio, etc.).

Let us note that a measure analogous to the ones of [1.2] or, more generally, of [1.3] can be obtained also for the reoptimization approach. Consider a probabilistic combinatorial optimization graph-problem $\text{P}\Pi$, derived from an optimization graph-problem Π and let $G(V, E)$ be a generic instance for the latter problem. Set $n = |V|$

and consider a vector $(p_1, \ldots, p_n)$ of presence-probabilities on the vertices of V. Then the functional $E^*(G)$ of the reoptimization for PΠ is defined as:

$$E^*(G) = \sum_{V' \subseteq V} m\left(G\left[V'\right], S^*\left(V'\right)\right) \Pr\left[V'\right] \qquad [1.4]$$

where $S^*(V')$ is an optimal solution for Π in $G[V']$, and $\Pr[V']$ is as in [1.1].

1.3. The main methodological issues dealing with probabilistic combinatorial optimization

1.3.1. *Complexity issues*

1.3.1.1. *Membership in **NPO** is not always obvious*

As one can see from [1.3] computation of functional's value is not *a priori* polynomial, since this expectation carries over all the possible subsets of the initial data-set. So, with respect to Definition 1.1, probabilistic versions of **NPO** problems do not trivially belong to **NPO** too. As we will see in the next chapters, when dealing with strategy MS sketched at the end of section 1.2, we succeed by more or less simple algebraic manipulations to show that functionals associated with it can be polynomially computed. This is the case for the problems dealt with in the next chapters as well as for the problems studied in [AVE 95, AVE 94, BEL 93, BER 88, BER 89, BER 90a, BER 90b, JAI 85, JAI 88a, JAI 92, JAI 88b, SÉG 93]. The basic idea underlying such a simplification is the following: instead of computing the value of the solution induced by any subinstance (recall that there exist an exponential number of subinstances of a given initial instance), one tries to determine, for any element of the *a priori* solution, the number of subinstances for which this element remains part of this solution. Even if this simplification technique works for numerous problems (associated with strategy MS), we will see in Chapters 2 and 3 that it quickly attains its limits once one tries to enrich MS with elementary operations improving its result. In particular, we will see that matching MS with natural greedy improvement techniques largely complicates the corresponding functionals in such a way that it is not obvious that their computation can be performed in polynomial time.

1.3.1.2. *Complexity of deterministic vs. complexity of probabilistic optimization problems*

Obviously, for any probabilistic combinatorial optimization graph-problem PΠ defined on a graph $G(V, E)$, if $p_i = 1$, for any $v_i \in V$, then PΠ coincides with Π in the sense that for any *a priori* solution S for PΠ, its functional has the same value as the objective value of S seen as solution of Π. This remark implies that if the functional[6] is computable in polynomial time and if Π is **NP**-hard, then PΠ, being a generalization of Π, is also **NP**-hard. Conversely, if Π is polynomial, then no immediate result can be deduced for PΠ.

Consider for instance two classical polynomial problems, the shortest path problem for fixed departure- and arrival-vertices s and t, respectively, and the minimum spanning tree problem. A probabilistic version for the former one is defined and studied in [JAI 92]. There, the input graph is complete, its vertices are independent and uniformly distributed in the unit square, some vertices are always present (i.e., they have probability 1), in particular s and t, and the rest of the vertices all have the same presence probability. Given an *a priori* solution, the adopted modification strategy consists of removing the absent vertices from this solution (this is not a problem since the input graph is assumed complete). As it is proved in [JAI 92], this version of probabilistic shortest path problem is **NP**-hard. The same holds for the minimum spanning tree problem ([BER 88, BER 90a]). For this problem, the input is the same as for shortest path. The modification strategy considered is the following: given an *a priori* tree T and a subgraph $G[V']$ of the input-graph, we consider the subtree of T restricted to the vertices of V' together with some vertices of $V \setminus V'$ (and the edges of T incident to these vertices) in order to guarantee connectivity of the induced subtree. This probabilistic version of minimum spanning tree is **NP**-hard.

When the deterministic version Π of a probabilistic problem PΠ is **NP**-hard, an interesting mathematical problem is to determine the complexity of PΠ for the classes of instances where Π is polynomial. Here also, results for the probabilistic problem are, very frequently, opposite to the ones for its (deterministic) support. For instance, as we will see in Chapter 5, the probabilistic versions of many coloring problems studied there are **NP**-hard, even for graph-classes for which deterministic coloring is polynomial.

Another interesting fact that will be clear in the next chapters (mainly in Chapters 4 and 5) is the role that the specific probability-system considered plays in complexity or approximation behaviors of the problems dealt. For instance, the fact that one assumes identical or distinct vertex-probabilities can completely change the complexity of a problem or its approximability.

Notice that an analogous fact can be established for the probabilistic traveling salesman, even when the input-graph K_n (the complete graph on n vertices) has identical vertex-probabilities. Denote by T^* an optimal *a priori* tour in K_n (i.e., an optimal

6. Recall that a probabilistic combinatorial optimization problem is always defined (see Definition 1.2) with respect to some modification strategy `M` that strongly affects the mathematical expression of its functional; for simplicity, when no confusion arises, this fact will be omitted.

solution of probabilistic traveling salesman under MS) and by T_0^* an optimal tour for the deterministic counterpart. In [JAI 85], counter-examples are given showing that $T^* \neq T_0^*$. In [BEL 93], it is shown that if n is odd ($n = 2k + 1$), then:

$$E\left(K_n, T^*, \mathtt{MS}\right) \geqslant p^2 m\left(K_n, T^*\right) \frac{1-(1-p)^{n-1}}{1-(1-p)^k} \qquad [1.5]$$

From [1.5], one can deduce the following two estimations:

$$\frac{m\left(K_n, T^*\right) - m\left(K_n, T_0^*\right)}{m\left(K_n, T_0^*\right)} \leqslant \frac{1-p^2}{p^2} \qquad [1.6]$$

$$\frac{E\left(K_n, T_0^*, \mathtt{MS}\right) - E\left(K_n, T^*, \mathtt{MS}\right)}{E\left(K_n, T^*, \mathtt{MS}\right)} \leqslant \frac{1-p^2}{p^2} \qquad [1.7]$$

The bounds given in [1.6] and [1.7] show that T_0^* constitutes a good approximation for T^* only in the case where p is large, i.e., when the probabilistic version becomes "close" to the deterministic one.

1.3.2. *Solution issues*

As we have already mentioned, the solution of probabilistic problems is not trivially deduced from the one of their deterministic (original) counterpart. In particular, optimal solution of the latter is very bad for the former.

This is, for example, the case for probabilistic traveling salesman in [JAI 85]. This is also the case for probabilistic coloring in bipartite graphs as mentioned above at the end of section 1.3.1.2. An absolutely vital step to take in order to solve a probabilistic combinatorial optimization problem is the characterization of its optimal *a priori* solution. This, as we will see later, is not always trivial for some modification strategies. Then, based upon this characterization, one can try to estimate the complexity of computing this solution.

1.3.2.1. *Characterization of optimal a priori solutions*

For numerous combinatorial optimization problems, it is possible to characterize the optimal *a priori* solution in terms of parameters of the initial input-graph. For example, as we will see in Chapters 2, 3 and 4, under MS, the optimal *a priori* solution for the problems covered there (probabilistic independent set, probabilistic vertex covering and probabilistic longest path where the solution is measured according to the number of its vertices, respectively) is the optimal solution of the corresponding weighted problem where vertex-weights are the corresponding probabilities. Then the complexity of the probabilistic problem is the same as the complexity of its weighted deterministic counterpart.

But there exist, conversely, functionals (always associated with MS) for which precise characterization of the optimal *a priori* solution in terms of input-graph parameters is not possible. This is mainly due to the fact that the weight of a vertex (seen as function of its probability) depends on the *a priori* solution itself and, consequently, it cannot be independently defined as a parameter depending only on the structure of the input-graph. For example, as we will see in Chapter 4, under the modification strategy MS, the functional of the probabilistic longest path (in transitive graphs) where the solution is measured with respect to the sum of the weights on its arcs and for an *a priori* solution $S = (0, 1, \dots, k, k+1)$ (where $0, \dots, k+1$ are the vertices of the *a priori* path) is expressed as:

$$\begin{aligned} E(G, S, \mathtt{MS}) &= \sum_{i=0}^{k} p_i p_{i+1} d(i, i+1) \\ &+ \sum_{i=0}^{k-1} \sum_{j=i+2}^{k+1} p_i p_j \left(\prod_{l=i+1}^{j-1} (1 - p_l) \right) d(i, j) \end{aligned} \quad [1.8]$$

where $d(i, j)$ is the weight (distance) of arc (i, j). As one can see from [1.8], if one tries to express this probabilistic problem in terms of some weighted version of its support, then one has to assign distance $p_i p_j d(i, j)$ to an arc (i, j) (of distance $d(i, j)$ in the initial graph) if it belongs to the *a priori* solution; otherwise, the distance of (i, j) would be equal to $p_i p_j (\prod_{l=i+1}^{j-1} (1-p_l)) d(i, j)$. This latter distance depends on S since it takes into account probabilities of vertices l lying between i and j in S. A corollary of this fact is that although the longest path problem is polynomial in transitive directed acyclic graphs (dags), this result does not hold (until now) for its probabilistic counterpart just discussed.

The same phenomenon appears for the probabilistic shortest path (under MS), considered in section 1.3.1.2 of Chapter 4. Number, arbitrarily, the vertices of K_n (recall that the input-graph is assumed to be complete and that some vertices are always present, i.e., they have probabilities equal to 1) and let a path $S = (0, 1, \dots, k+1)$ from s to t (i.e., s is numbered by 0 and t by $k+1$) be an *a priori* solution for probabilistic shortest path in K_n. Let $d_S(i, i+r+1) = \sum_{e=0}^{s} d(b_e, b_{e+1})$, where $b_0 = i$, $b_{s+1} = i + r + 1$ and $(b_1, \dots, b_s)$ is the sequence of vertices always present between $(i+1, \dots, i+r)$; let W be the number of present vertices between the ones having a presence probability p different from 1. As it is shown in [JAI 92]:

$$E\left(K_n, S, \texttt{MS}\right) =$$

$$\sum_{r=0}^{k-2}\sum_{j=r}^{n-2} \frac{\binom{n-2-r}{j-r}}{\binom{n}{j}} \Pr(W = n - j) \sum_{i=1}^{k-1-r} d_S(i, i + r + 1)$$

$$+ \sum_{r=0}^{k-1}\sum_{j=r}^{n-1} \frac{\binom{n-1-r}{j-r}}{\binom{n}{j}} \Pr(W = n - j) \left(d_S(0, r + 1) + d_S(k - r, k + 1)\right)$$

$$+ \sum_{j=k}^{n} \frac{\binom{n-k}{j-k}}{\binom{n}{j}} \Pr(W = n - j) d_S(0, k + 1) \qquad [1.9]$$

As we can see in [1.9], the distances one could assign to the arcs strongly depend on the *a priori* solution and, consequently, no precise characterization of this solution is possible.

1.3.2.2. *Polynomial subcases*

The identification of polynomial restrictive cases for **NP**-hard problems is always an interesting issue in complexity theory. It is also the case for probabilistic combinatorial optimization. The most common approach for such an issue is to start from polynomial instances for the deterministic support and to study if property guaranteeing polynomial solution there remains valid for the probabilistic counterpart.

Consider, for instance, the traveling salesman problem and its probabilistic version under MS. As it is shown in [BER 79], matrices of the form $c_{ij} = a_i + b_j$ (called *constant* matrices) are the only ones where all the permutations of vertices have the same length. Based upon this result, it is shown in [BER 88] (see also [BEL 93]) that the constant matrices are the only ones that have the same expectation for any *a priori* tour T and this expectation is equal to $p(1 - (1 - p)^{n-1})m(K_n, T)$. So, in the case of constant matrices, the probabilistic traveling salesman is polynomial under identical vertex-probabilities.

Let us give another example of polynomial subcases always dealing with the probabilistic traveling salesman. Call a matrix C *small*, if there exist two vertex-vectors, $\vec{a}$ and $\vec{b}$ such that $c_{ij} = \min\{a_i, b_j\}$, $i, j = 1, \dots, n$. A small matrix where a_i's and b_j's are all distinct is called *small with distinct values*. In this case, let d_i be the i-th smallest value between the $2n$ values a_k and b_j. Let $D = \{d_1, \dots, d_n\}$, $\bar{D} = \{d_{n+1}, \dots, d_{2n}\}$ and $d = \sum_{i=1}^{n} d_i$. We set $D_2 = \{i : \{a_i, b_i\} \subseteq D\}$, $D_0 = \{i : \{a_i, b_i\} \subseteq \bar{D}\}$, $D_a = \{i : a_i \in D, b_i \in \bar{D}\}$ and $D_b = \{i : b_i \in D, a_i \in \bar{D}\}$. As it is shown in [LAW 85], the traveling salesman problem is polynomial when dealing with small matrices. In particular, the value of the optimal tour is equal to d if and only if D satisfies one of the following conditions:

1) $D_2 \neq \emptyset$;

2) $D = \{a_1, \ldots, a_n\}$;

3) $D = \{b_1, \ldots, b_n\}$.

Otherwise, the value of the optimal tour equals $d' = d - d_n + d_{n+1}$, if and only if $D' = D \cup \{d_{n+1}\} \setminus \{d_n\}$ satisfies one of the following conditions:

a) $D'_2 \neq \emptyset$, where D'_2 is defined analogously to D_2;

b) $D' = \{a_1, \ldots, a_n\}$;

c) $D' = \{b_1, \ldots, b_n\}$.

Based upon the above, it is proved in [BEL 93] that, in the case of identical vertex-probabilities, if C is a small matrix with distinct values, then, setting $q = 1 - p$:

– $E(K_n, T^*, \mathtt{MS}) = p(1 - q^{n-1})d$, if and only if conditions 2 and 3 are verified;

– if conditions 1, 2 and 3 are not verified then:

- $E(K_n, T^*, \mathtt{MS}) = p(1 - q^{n-1})d'$, if and only if D' verifies conditions a) and b);

- $E(K_n, T^*, \mathtt{MS}) = p(1 - q^{n-1}) \min\{d + d_{n+2} - d_n, d - d_{n-1} + d_{n+1}\}$, if and only if D' does not verify conditions a), b) and c) and either one of the following conditions (c1), (c2) is satisfied: (c1) $D \cup \{d_{n+2}\} \setminus \{d_n\}$ satisfies condition b) or c) and $d + d_{n+2} - d_n \leqslant d - d_{n-1} + d_{n+1}$; (c2) $D \cup \{d_{n+1}\} \setminus \{d_{n-1}\}$ satisfies condition b) or c) and $d + d_{n+2} - d_n \geqslant d - d_{n-1} + d_{n+1}$.

If one of the two basic items above is satisfied, then $T^* = T_0^*$ and, following the results of [LAW 85], the probabilistic traveling salesman is polynomial.

In the next chapters of this book, several polynomial subcases are given for the problems covered. Let us note that, in general, when optimal *a priori* solutions of probabilistic problems coincide with optimal solutions of their deterministic supports, then *a priori* optimization coincides with reoptimization.

1.3.2.3. *Exact solutions and polynomial approximation issues*

Whenever problems considered are **NP**-hard, or they cannot be proved polynomial, then they can be obviously solved by optimal algorithms even if these algorithms are exponential. In Chapter 5 we present such algorithms for probabilistic coloring in restrictive cases of bipartite graphs as trees and chains. But, dealing with effective solution of probabilistic combinatorial optimization problems, this monograph focuses on polynomial approximation of the problems studied.

In general, there exist three types of polynomial approximation results obtained for a probabilistic combinatorial optimization problem:

1) one measures, for a given optimization problem, the quality of the *a priori* optimization with respect to the reoptimization; for some modification strategy M, this can be done by means of the ratio $E(G, S, \mathtt{M})/E^*(G)$, where $E(G, S, \mathtt{M})$ and $E^*(G)$ are given by [1.2] and by [1.4], respectively;

2) one measures the quality of a solution S obtained in the (deterministic) support and without taking into account any probabilistic concept, when used as an *a priori* solution for the probabilistic counterpart (for a fixed modification strategy M); this is done by means of the ratio $E(G, S, \mathtt{M})/E(G, S^*, \mathtt{M})$, where S^* is the optimal *a priori* solution (associated with M) and both $E(G, S, \mathtt{M})$ and $E(G, S^*, \mathtt{M})$ are given by [1.2];

3) finally, one measures the quality of an *a priori* solution S, explicitly built for the probabilistic problem[7]; this quality is measured by the ratio of item 2.

For item 1, the interested reader can be referred to [JAI 93], which deals with the probabilistic traveling salesman and probabilistic minimum spanning tree, both problems defined on complete graphs with identical vertex probabilities and with vertices uniformly distributed on $\mathbb{R}^2$, under strategy MS.

For item 2, that is somewhat closer to the spirit of this book than item 1, we quote the study performed in [BER 88]. There, dealing with probabilistic traveling salesman under the same assumptions as in [JAI 93] also, the *a priori* tour T considered is the one computed by the celebrated Christofides' algorithm ([CHR 76]). This algorithm, based upon a minimum spanning tree computation on K_n, achieves an approximation ratio 2 for the traveling salesman problem in metric spaces. It is shown in [BER 88] that, denoting by T^*, an optimal *a priori* tour (i.e., an optimal *a priori* solution), then:

$$\frac{E\left(K_n, T, \mathtt{MS}\right)}{E\left(K_n, T^*, \mathtt{MS}\right)} \leqslant 2D$$

where D is the diameter of the minimum spanning tree intermediately computed by the Christofides' algorithm.

This result has been improved in [BEL 93] (under the same assumptions for the input graph and always under MS). It is proved that if X is a random variable representing the number of present vertices and verifying $\Pr(X \leqslant n-k-1) = 0$ and $\Pr(X = n-k) > 0$, then:

$$\frac{E\left(K_n, T, \mathtt{MS}\right)}{E\left(K_n, T^*, \mathtt{MS}\right)} \leqslant \frac{3}{2}\left(1 + \frac{k^2(k+1)}{n-2}\right) \qquad [1.10]$$

7. In other words, S is constructed by an algorithm taking more or less into account the probabilistic nature of the problem covered.

As one can see from [1.10], the ratio achieved is constant and tends to 3/2 for $n \rightarrow \infty$.

Approximation results presented in next chapters deal with item 3. For this reason, no example for this item is presented in this chapter.

1.4. Miscellaneous and bibliographic notes

In order to characterize the optimal *a priori* solution for a probabilistic combinatorial optimization problem, one has to rewrite the associated functional in an explicit (and hence intuitive) way. Whenever this is not possible, a statement about the complexity of its computation is impossible too. In this case, an interesting approach is to compute explicit (and non-trivial) bounds for it. The same holds for reoptimization. A complementary issue is the study of the asymptotic behavior for both *a priori* and reoptimization approaches. This is done in [BER 90b] for the probabilistic traveling salesman problem, the probabilistic minimum spanning tree problem, the probabilistic Steiner tree problem[8], the probabilistic vehicle routing problem[9] and the probabilistic facility location problem[10], under the assumptions that input-graphs are complete, vertex probabilities are identical ($p_i = p$, for any $v_i \in V$) and vertices are uniformly distributed in $\mathbb{R}^2$.

The result of [BER 90b] dealing with the asymptotic analysis of reoptimization is the following. Let κ_Π, Π standing for traveling salesman, minimum spanning tree and minimum Steiner tree, be constants ([STA 79, HAI 85]) for which, with probability 1, $\lim_{n\rightarrow\infty} m(K_n, T)/\sqrt{n} = \kappa_\Pi$, where T is some feasible solution for Π. Then, with probability 1:

$$\lim_{n\rightarrow\infty} \frac{E^*_{\text{P}\Pi}(K_n)}{\sqrt{n}} = \kappa_\Pi \sqrt{p}$$

8. Given a connected graph $G(V, E)$, a length function d on its edges, and a set $N \subseteq V$, the objective is to determine an optimal Steiner tree, i.e., a minimum-length tree spanning all vertices in N (the length of a tree is given by $d(T) = \sum_{e\in E(T)} d(e)$).

9. The general version of this problem is defined as follows: we are given a graph $G(V, E)$ together with a length $l(e)$, for any $e \in E$, a subset $E' \subseteq E$ and a subset $V' \subseteq V$; the objective is to determine a minimum-length cycle of G that visits any vertex in V' exactly once and traverses any edge in E', the length of such a cycle being the sum of the lengths of its edges.

10. Given a complete graph K_n on a set V of n vertices, costs $c(v_i, v_j)$, $v_i, v_j \in V$, that are symmetric and satisfy the triangle inequality, a set $F \subseteq V$ of locations where a facility may be built, a non-negative cost $f(v)$, $v \in V$, of building a facility at v and, for any location v, a non-negative demand $d(v)$, the objective is to determine a set $F' \subseteq F$ minimizing the quantity $\sum_{v\in F'} f(v) + \sum_{u\in F'} \sum_{v\in V} d(u)c(u, v)$.

Furthermore, if $E[r]$ is the expected radial distance from the depot[11] to a vertex of the input-graph for the vehicle routing problem and $C(n)$ is the vehicle capacity, then, with probability 1, the following holds:

$$\begin{array}{lll} \lim\limits_{n\to\infty} \frac{C(n)E^*(K_n)}{n} = 2E[r]p & \text{if } C(n) = o\left(\sqrt{n}\right) \\ \lim\limits_{n\to\infty} \frac{E^*(K_n)}{\sqrt{n}} = \kappa\sqrt{p} & \text{if } C(n) = \Omega\left(\sqrt{n}\right) \end{array}$$

where k is the constant of [STA 79, HAI 85] for the traveling salesman.

Finally, for any vertex i, the following holds, with probability 1, for the probabilistic facility location problem:

$$\lim_{n\to\infty} \frac{E_i^*(K_n)}{\sqrt{n}} = \kappa\sqrt{p}$$

where $E_i^*(K_n)$, is the functional of the reoptimization approach when the server's location vertex is i (the functional associated with the reoptimization depends, for the probabilistic facility location problem, on this vertex).

For the *a priori* optimization approach, it is proved in [BER 90b] that there exist quantities $c_{\mathrm{P}\Pi'}(p)$, Π' standing for traveling salesman and minimum spanning tree, such that, when dealing with an optimal *a priori* solution S^*, then with probability 1:

$$\lim_{n\to\infty} \frac{E_{\mathrm{P}\Pi'}(K_n, S^*, \mathtt{MS})}{\sqrt{n}} = c_{\mathrm{P}\Pi'}(p)$$

Furthermore, for probabilistic vehicle routing, the following hold with probability 1:

$$\begin{array}{lll} \lim\limits_{n\to\infty} \frac{C(n)E(K_n,S^*,\mathtt{M1})}{n} = 2E[r]p & \text{if } C(n) = o\left(\sqrt{n}\right) \\ \lim\limits_{n\to\infty} \frac{E(K_n,S^*,\mathtt{M1})}{\sqrt{n}} = \kappa_{\mathrm{P}\Pi} & \text{if } C(n) = \Omega\left(\sqrt{n}\right) \\ \lim\limits_{n\to\infty} \frac{C(n)E(K_n,S^*,\mathtt{M2})}{n} = 2E[r]p & \text{if } C(n) = o\left(\sqrt{n}\right) \\ \lim\limits_{n\to\infty} \frac{E(K_n,S^*,\mathtt{M2})}{\sqrt{n}} = c_{\mathrm{P}\Pi'}(p) & \text{if } C(n) = \Omega\left(\sqrt{n}\right) \end{array} \quad [1.11]$$

where Π and Π' in the subindices of κ and c in the second and fourth line of [1.11], respectively, stand both for the traveling salesman, $C(n)$ and $E[r]$ are as above and M1 and M2 are the modification strategies for probabilistic vehicle routing discussed in example 1.5 (page 20).

11. Located to the point $(0, 0)$ of $\mathbb{R}^2$.

Finally, for probabilistic facility location problem and for any i, the following hold, with probability 1:

$$\lim_{n \to \infty} \frac{E\left(K_n, S^*, \mathtt{MS}\right)(i)}{\sqrt{n}} = c_{\mathrm{P}\Pi'}(p)$$

where Π' in the subindex of c stands always for the traveling salesman.

Let us note that in [BER 90b], it is conjectured that:
– for Π standing for the traveling salesman, $c_{\mathrm{P}\Pi}(p) = \kappa_{\Pi}\sqrt{p}$;
– for Π standing for the minimum spanning tree, $c_{\mathrm{P}\Pi}(p) = \kappa_{\mathrm{P}\Pi}\sqrt{p}$.

Furthermore, [JAI 85], establishes the following bounds for $c_{\mathrm{P}\Pi}(p)$, Π standing for the traveling salesman: $\kappa_{\Pi}\sqrt{p} \leqslant c_{\mathrm{P}\Pi}(p) \leqslant \min\{\kappa_{\Pi}, 0.92\sqrt{p}\}$.

Another very interesting issue, covered only marginally in the literature until now, is the study of conditions under which the *a priori* approach and the reoptimization one are equivalent, i.e., identifying classes of instances for which $S^* = S_0^*$, where S_0^* and S^* are the optimal solution of the deterministic problem and the optimal *a priori* solution of its probabilistic derivation (under some modification strategy M). This equivalence induces a kind of solution's stability in the sense that if, following M, we modify S_0^* to fit the present subinstance of the problem at hand, then the solution so obtained is optimal for this subinstance.

The interested reader can find in [BEL 93] some interesting results concerning the probabilistic traveling salesman under MS. For this problem, equivalence between the *a priori* approach and the reoptimization approach means that if V' is the set of present vertices, then the tour induced by removing the $V \setminus V'$ absent ones from T_0^* is optimal for the graph $K_n[V']$.

Revisit the results about the traveling salesman presented in section 1.3.2.2. Based upon [LAW 85], one can construct an optimal solution for the traveling salesman in a small matrix. Then, since any submatrix of a small matrix is a small matrix itself, it is possible to construct optimal solutions for traveling salesman and for any vertex-subset V' of the initial input-graph. So, one can verify if, for any V', the tour induced by T_0^* is optimal for $K_n[V']$ or not. Consider, for example, the case where condition 1 is verified and, moreover, $|D_2| = 1$ and $D_b = \emptyset$. Since $|D_2| = |D_0|$, one can suppose $D_2 = \{1\}$ and $D_0 = \{n\}$. Under these assumptions, the following is shown in [BEL 93]. Let C be a small matrix. Then $T_0^* = T^*$, if and only if $((d_n = b_1) \vee ((d_n = a_1) \wedge (d_{n-1} = b_1))) \wedge ((d_{n+1} = a_n) \vee ((d_{n+1} = b_n) \wedge (d_{n+2} = a_n)))$. In this case:

$$E\left(K_n, T^*, \mathtt{MS}\right) = p\left(1 - (1-p)^{n-1}\right)d' - p^2\left(d_{n+1} - d_n\right)$$

Furthermore, let C be a small matrix and consider the following conditions:

1) $(d_n \neq b_1) \wedge ((d_n \neq a_1) \vee (d_{n-1} \neq b_1)) \wedge ((d_{n+1} = a_n) \vee ((d_{n+1} = b_n) \wedge (d_{n+2} = a_n)))$;

2) $((d_n = b_1) \vee ((d_n = a_1) \wedge (d_{n-1} = b_1))) \wedge (d_{n+1} \neq a_n) \wedge ((d_{n+1} \neq b_n) \vee (d_{n+2} \neq a_n))$.

Then, setting $q = 1 - p$, if 1 above is verified, we get:

$$E(K_n, T^*, \mathtt{MS}) = p\left(1 - q^{n-1}\right)(d + a_n) - p^2 \sum_{r=0}^{n-2} q^r \max\{a_{n-r}, b_1\}$$

On the other hand, if 2 is verified, then:

$$E(K_n, T^*, \mathtt{MS}) = p\left(1 - q^{n-1}\right)(d - b_1) + p^2 \sum_{r=0}^{n-2} q^r \min\{a_n, b_{1+r}\}$$

We conclude this section with an approximation result that cannot be classified in the personal classification, about approximation results which has been proposed in section 1.3.2.3.

In [BER 88], the following is shown. If $m(K_n, S_0^*)$ is the optimal solution value for a deterministic graph-problem Π, if $m(K_n, S)$ is the value of the solution computed by an approximation algorithm assumed to solve Π, if, for an instance G of Π, p is the presence probability of its vertices, if Π stands for the minimum traveling salesman, the minimum spanning tree and the vehicle routing problem and if:

$$\frac{m(K_n, S)}{m(K_n, S_0^*)} \leqslant \rho$$

(all the three problems represented by Π are minimization ones), then for any modification strategy $\mathtt{M}$, and under the assumptions on the instances met throughout this chapter (graphs are complete and vertices are uniformly distributed in $\mathbb{R}^2$):

$$E\left[\frac{m(K_n, S_0^*)}{E(K_n, S_0^*, \mathtt{M})}\right] \leqslant \frac{\rho}{p}$$

FIRST PART

Probabilistic Graph-problems

Chapter 2

The Probabilistic Maximum Independent Set

In this chapter, we study the complexity of optimally solving probabilistic maximum independent set problem using several *a priori* optimization strategies, as well as the complexity of approximating optimal solutions.

An instance of PROBABILISTIC MAX INDEPENDENT SET is a pair $(G, \vec{\Pr})$ and is obtained by associating with each $v_i \in V$ an "occurrence" probability p_i and by considering a modification strategy $\mathtt{M}$ transforming a feasible independent set S of G into an independent set for the subgraph of G induced by a set $V' \subseteq V$. The objective for PROBABILISTIC MAX INDEPENDENT SET is to determine the *a priori* solution $\hat{S}$ maximizing the functional $E(G, S, \mathtt{M})$ defined as (definition 1.2 in Chapter 1) $\sum_{V' \subseteq V} m(V', S(V', \mathtt{M})) \Pr[V']$.

Except for its theoretical interest, PROBABILISTIC MAX INDEPENDENT SET has also concrete applications. In [GAB 97], some aspects of the satellite shots planning problem are studied. A graph-theoretic modelling for this problem is proposed there and it is proved that, via this modelling, the solution of the problem studied becomes the computation of a maximum independent set in a type of graph called a "conflict graph" in which a vertex represents a shot to be realized. However, it is not taken into account that shots realized under strong cloud-covering are not operational. Consequently, in order to compute an exploitable an operational solution, it is essential to also model weather forecasting. This can be done by associating probability p_i with vertex v_i of the conflict graph; the higher the vertex-probability, the more operational the shot taken. In this way, we naturally obtain a model leading to a probabilistic combinatorial optimization problem. Such a model for the satellite shots planning problem allows, given an *a priori* MAX INDEPENDENT SET-solution, computation of the expected number of operational shots.

There exist two interpretations of such an approach, each one characterized by its proper modification strategy:

– the plan is firstly executed and one can know only after the plan's execution if a shot is operational; in this case, one retains only the operational ones among the shots realized; this, in terms of PROBABILISTIC MAX INDEPENDENT SET, amounts to an application of strategy M1 introduced in section 2.1;

– weather forecasting becomes a certitude just before the plan's execution; in this case, starting from an *a priori* MAX INDEPENDENT SET-solution, one knows the vertices of this solution corresponding to non-operational shots, one discards them from the *a priori* solution and, finally, one renders the survived solution maximal by completing it by new vertices corresponding to operational shots; this amounts to application of other strategies, for example the ones denoted by M2, M3, M4, or M5 in the sequel and introduced in section 2.1.

Let us note that the probabilistic extension of the model of [GAB 97] can also be used to represent another concept, modelled in terms of PROBABILISTIC MAX INDEPENDENT SET, where randomness on vertices this time represents probabilities that the corresponding shots are requested. Shot-probability equal to 1 means that this shot has already been requested, while shot-probability in [0,1] means that the corresponding shot will eventually be requested just before its realization. The corresponding PROBABILISTIC MAX INDEPENDENT SET can be effectively solved by applying strategy M2 ([MUR 97]), M3, M4, or M5.

In what follows we consider maximal *a priori* independent sets and use five modification strategies, $\mathtt{Mi}$, $\mathtt{i} = 1, \ldots, 5$. For M1 and M5 we express their functionals in a closed form, we prove that they are computed in polynomial time, and we determine the *a priori* solutions that maximize them. For M2 and M3, the expressions for the functionals are more complicated and it seems that they cannot be computed in polynomial time. Due to the complicated expressions for these functionals, we have not been able to characterize the *a priori* solutions maximizing them. Finally, for M4, we prove that the functional associated can be computed in polynomial time, but we are not able to precisely characterize the optimal *a priori* solution maximizing it. For all the strategies proposed we also study the complexity of approximating optimal *a priori* solutions.

We recall here that the strategies studied in fact introduce *five distinct* probabilistic combinatorial optimization problems denoted in the sequel by PROBABILISTIC MAX INDEPENDENT SET1, PROBABILISTIC MAX INDEPENDENT SET2, PROBABILISTIC MAX INDEPENDENT SET3, PROBABILISTIC MAX INDEPENDENT SET4 and PROBABILISTIC MAX INDEPENDENT SET5, respectively. Finally, we study the probabilistic version of a natural restriction of MAX INDEPENDENT SET, the one where the input graph is bipartite.

In this chapter, given a graph $G(V, E)$ of order n, we sometimes denote by $V(G)$ the vertex-set of G. We denote by S a maximal solution of MAX INDEPENDENT SET of G, by S^* a maximum independent set of G, by $\alpha(G)$ its cardinality (see section A.2 of Appendix A) and by $\hat{S}$ an optimal PROBABILISTIC MAX INDEPENDENT SET-solution (optimal *a priori* solution). By $\Gamma(V')$, $V' \subseteq V$, we denote the set $\cup_{v_i \in V'} \Gamma(v_i)$; by $\Pr[v_i] = p_i$, we denote the fact that the presence probability of a vertex $v_i \in V$ equals p_i. As adopted in Appendix A (section A.2), given a set $V' \subseteq V$, we denote by $G[V'](V', E'_V)$ the subgraph of G induced by V' (obviously, there are 2^n such graphs). Given a maximal solution S of MAX INDEPENDENT SET (the *a priori* solution) in G, we denote by $S(V')$ the set $S \cap V'$

2.1. The modification strategies and a preliminary result

In what follows we denote by GREEDY the classical greedy MAX INDEPENDENT SET-algorithm. It works as follows:

1) set $S = \emptyset$;

2) order the vertices of V in non-decreasing degree-order;

3) include in S a minimum-degree vertex v_0 of G;

4) delete $\{v_0\} \cup \Gamma(v_0)$ from G together with any edge incident to these vertices;

5) repeat Steps 2 to 3 until all vertices are removed;

6) output S.

Moreover, we denote by SIMGREEDY a simplified version of GREEDY where after removing a vertex and its neighbors, the algorithm does not reorder the vertices of the surviving graph, i.e., it does not re-execute Step 2.

2.1.1. *Strategy* M1

Given an *a priori* MAX INDEPENDENT SET-solution S and a present subset $V' \subseteq V$, modification strategy M1 consists of simply moving the absent vertices out of S, i.e., of considering set $S(V') = S'_1$ as solution for $G[V']$. Observe that M1 is, for PROBABILISTIC MAX INDEPENDENT SET, the strategy denoted by MS in Chapter 1.

EXAMPLE 2.1.– Consider the graph of Figure 2.1 and the *a priori* independent set $S = \{1, 3, 7, 8\}$. Assuming that vertices 3 and 8 are absent, application of strategy M1 on the surviving graph produces as solution the set $\{1, 7\}$ (Figure 2.2).

2.1.2. *Strategies* M2 *and* M3

Modification strategy M2 is a two-step method: it first applies M1 to obtain $S(V')$; next, it applies GREEDY on the graph $G[\tilde{V}'] = G[V' \backslash \{S(V') \cup \Gamma(S(V'))\}]$ and, finally,

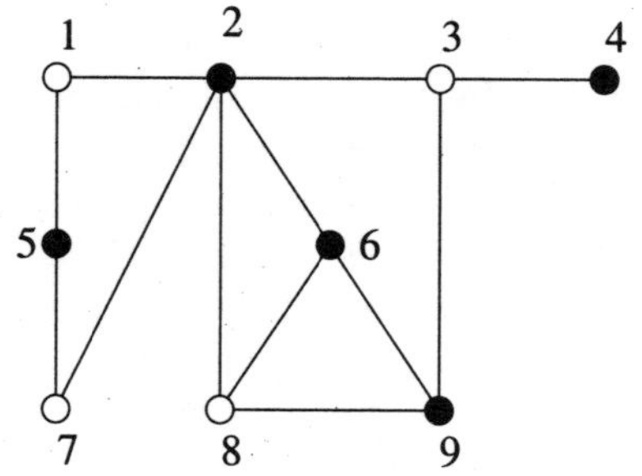

Figure 2.1. *A graph G, together with the a priori independent set* $\{1,3,7,8\}$ *("white" vertices)*

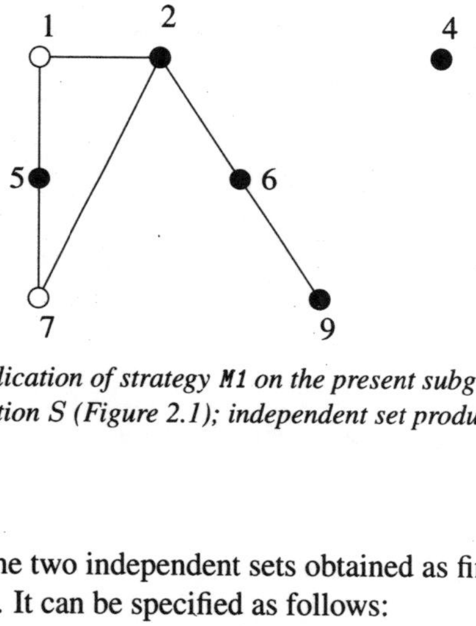

Figure 2.2. *Application of strategy* M1 *on the present subgraph of* G*, with a priori solution* S *(Figure 2.1); independent set produced:* $\{1,7\}$

it retains the union of the two independent sets obtained as final MAX INDEPENDENT SET-solution for $G[V']$. It can be specified as follows:

1) set: $S(V') = S(V', \texttt{M1})$, $\tilde{V}' = V' \setminus \{S(V') \cup \Gamma(S(V'))\}$;

2) set $S(\tilde{V}') = \texttt{GREEDY}(G[\tilde{V}'])$;

3) output $S(V', \texttt{M2}) = S_2' = S(V') \cup S(\tilde{V}')$.

Strategy M3 is identical to M2 modulo the fact that, instead of GREEDY, algorithm SIMGREEDY is executed at Step 2. The solution $S(V', \texttt{M3})$ computed by strategy M3 will be denoted by S_3'.

EXAMPLE 2.2.– Consider again the graph G and the *a priori* independent set S of Figure 2.1 (with absent vertex-set $\{3,8\}$). Application of strategy M2 or M3 will first produce solution $S(V') = \{1,7\}$ (Step 1). Then, application of Steps 2 and 3 will add vertices 4 and 6 to $S(V')$. Finally, $S_2' = \{1,4,6,7\}$.

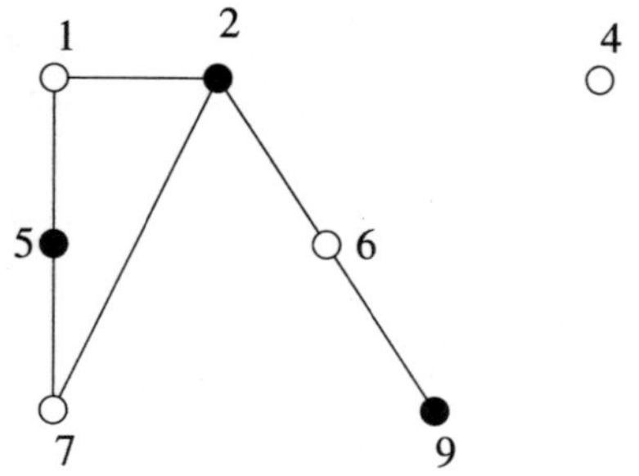

Figure 2.3. *Application of strategy M2 on the present subgraph of G, with a priori solution S (Figure 2.1); independent set produced:* $\{1, 4, 6, 7\}$

2.1.3. *Strategy* M4

Strategy M4 starts from $S(V')$ and completes it with the isolated vertices (vertices with no neighbors) of the graph $G[\tilde{V}']$. It is specified as follows:

– set: $S(V') = S(V', \texttt{M1})$, $\tilde{V}' = V' \setminus \{S(V') \cup \Gamma(S(V'))\}$;

– output $S(V', \texttt{M4}) = S'_4 = S(V') \cup \{v_i \in \tilde{V}' : \Gamma(v_i) = \emptyset\}$.

EXAMPLE 2.3.– Consider again the graph G and the *a priori* independent set S of Figure 2.1 with vertices 3 and 8 absent. Application of strategy M4 will first produce solution $S(V') = \{1, 7\}$ (first step). Then, application of the second step will add to $S(V')$ the isolated vertex 4. Finally, $S'_4 = \{1, 4, 7\}$.

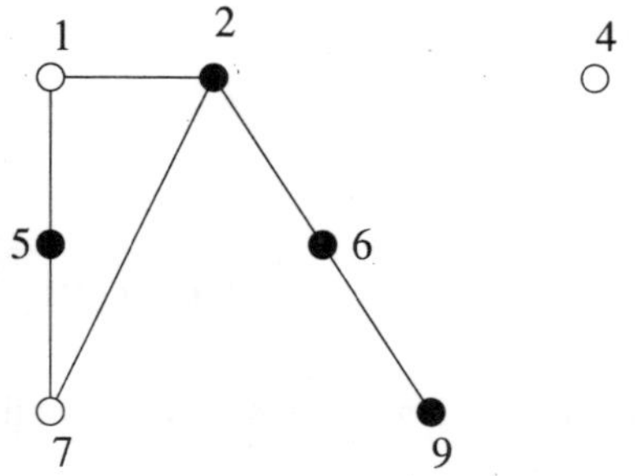

Figure 2.4. *Application of strategy M4 on the present subgraph of G, with a priori solution S (Figure 2.1); independent set produced:* $\{1, 4, 7\}$

2.1.4. *Strategy* M5

Strategy M5 applies the natural relation between a vertex cover and an independent set in a graph, mentioned in Appendix A (section A.2), that a vertex cover (resp., independent set) is the complement, with respect to the vertex set of the graph, of an independent set (resp., vertex cover). This strategy is specified as follows:

1) set: $C = V \setminus S$, $C(V') = C \cap V'$;

2) set $R = \{v_i \in C(V') : \Gamma(v_i) = \emptyset\}$;

3) set $C(V') = C(V') \setminus R$;

4) output $S(V', \mathtt{M5}) = S'_5 = V' \setminus C(V')$.

EXAMPLE 2.4.– Figure 2.5 illustrates the application of M5 in the graph G and the *a priori* independent set S of Figure 2.1, always supposing that vertices 3 and 8 are absent. Step 1 will produce $C(V') = \{2, 4, 5, 6, 9\}$. Step 2 will compute $R = \{4\}$, while Step 3 results in $C(V') = \{2, 5, 6, 9\}$. Finally, Step 4 will return $S'_5 = \{1, 4, 7\}$ (Figure 2.5).

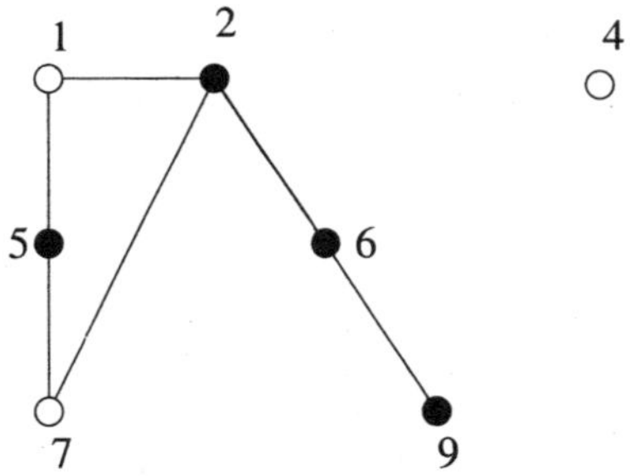

Figure 2.5. *Example of application of strategy* **M5** *on the present subgraph G, with a priori solution S (Figure 2.1); independent set produced:* $\{1, 4, 7\}$

2.1.5. *A general mathematical formulation for the five functionals*

We establish in this section a general expression valid for the functionals of all the five strategies studied in the chapter.

THEOREM 2.1.– *Consider an a priori solution S of cardinality $|S|$ for G; consider strategies* Mk, k = 1, ..., 5. *With each vertex $v_i \in V$ we associate a probability p_i and a random variable $X_i^{\mathtt{Mk},S}$,* k = 1, ..., 5, *defined, for any $V' \subseteq V$, by:*

$$X_i^{\mathtt{Mk},S} = \begin{cases} 1 & \text{if } v_i \in S'_k \\ 0 & \text{otherwise} \end{cases} \qquad [2.1]$$

Then:

$$E\left(G, S, \texttt{Mk}\right) = \sum_{v_i \in S} p_i + \sum_{v_i \in (V \setminus S)} \Pr\left[X_i^{\texttt{Mk},S} = 1\right] \qquad [2.2]$$

In particular, if, for each vertex $v_i \in V$, $p_i = p$, then:

$$E\left(G, S, \texttt{Mk}\right) = p|S| + \sum_{v_i \in (V \setminus S)} \Pr\left[X_i^{\texttt{Mk},S} = 1\right] \qquad [2.3]$$

Proof. By [2.1], $|S'_k| = \sum_{i=1}^{n} X_i^{\texttt{Mk},S}$. So:

$$\begin{aligned}
E\left(G, S, \texttt{Mk}\right) &= \sum_{V' \subseteq V} \Pr\left[V'\right] |S'_k| \\
&= \sum_{V' \subseteq V} \Pr\left[V'\right] \sum_{i=1}^{n} X_i^{\texttt{Mk},S} = \sum_{i=1}^{n} \sum_{V' \subseteq V} \Pr\left[V'\right] X_i^{\texttt{Mk},S} \\
&= \sum_{i=1}^{n} E\left(X_i^{\texttt{Mk},S}\right) = \sum_{i=1}^{n} \Pr\left[X_i^{\texttt{Mk},S} = 1\right] \\
&= \sum_{i=1}^{n} \Pr\left[X_i^{\texttt{Mk},S} = 1\right] \left(1_{\{v_i \in S\}} + 1_{\{v_i \notin S\}}\right) \\
&= \sum_{i=1}^{n} \Pr\left[X_i^{\texttt{Mk},S} = 1\right] 1_{\{v_i \in S\}} \\
&\quad + \sum_{i=1}^{n} \Pr\left[X_i^{\texttt{Mk},S} = 1\right] 1_{\{v_i \notin S\}}
\end{aligned}$$

But, if $v_i \in S$, then necessarily $X_i^{\texttt{Mk},S} = 1$, $\forall V'$, such that $v_i \in V'$; so, for each $v_i \in S$:

$$\Pr\left[X_i^{\texttt{Mk},S} = 1\right] = p_i$$

and consequently:

$$\begin{aligned}
E\left(G, S, \texttt{Mk}\right) &= \sum_{v_i \in S} p_i + \sum_{i=1}^{n} \Pr\left[X_i^{\texttt{Mk},S} = 1\right] 1_{\{v_i \notin S\}} \\
&= \sum_{v_i \in S} p_i + \sum_{v_i \in (V \setminus S)} \Pr\left[X_i^{\texttt{Mk},S} = 1\right]
\end{aligned}$$

If $p_i = p$, $v_i \in V$, then the result of [2.3] is immediately obtained from [2.2]. ∎

Let us note that, as it can be easily deduced from the proof of Theorem 2.1, the above result holds for any strategy which firstly determines $S(V')$, it next computes an independent set $S(\tilde{V}')$ on $G[\tilde{V}']$, and it finally considers as solution for $G[V']$ the set $S(V') \cup S(\tilde{V}')$.

2.2. PROBABILISTIC MAX INDEPENDENT SET1

2.2.1. *Computing optimal a priori solutions*

From [2.1], we have $X_i^{\mathtt{M1},S} = 0$, $\forall v_i \notin S$, and consequently, $\Pr[X_i^{\mathtt{M1},S} = 1] = 0$, for $v_i \in V \setminus S$; so, the following theorem is immediately derived for strategy M1.

THEOREM 2.2.– *Given a graph $G(V,E)$, an a priori solution S and the modification strategy* M1, *then $E(G,S,\mathtt{M1}) = \sum_{v_i \in S} p_i$, and is computed in $O(n)$. Optimal* PROBABILISTIC MAX INDEPENDENT SET*1-solution $\hat{S}$ is a maximum-weight independent set in a weighted version of G where vertices are weighted by the corresponding probabilities. If $p_i = p$, $\forall v_i \in V$, then $E(G,S,\mathtt{M1}) = p|S|$; in this case, $\hat{S} = S^*$ and $E(G,\hat{S},\mathtt{M1}) = p\alpha(G)$.*

The characterization of $\hat{S}$ given in Theorem 2.2 immediately introduces the following complexity result for PROBABILISTIC MAX INDEPENDENT SET1.

THEOREM 2.3.– PROBABILISTIC MAX INDEPENDENT SET*1 is **NP**-hard.*

We now show that for $p_i = p$, $v_i \in V$, a mathematical expression for $E(G,S,\mathtt{M1})$ can be built directly without applying Theorem 2.1 (used in next sections for the analysis of other strategies). Given that $0 \leqslant |S_1'| \leqslant |S|$, we get:

$$|S_1'| = |S(V')| \sum_{i=1}^{|S|} 1_{\{|S(V')|=i\}}$$

So, the functional for M1 can be written as:

$$\begin{aligned} E(G,S,\mathtt{M1}) &= \sum_{V' \subseteq V} \Pr[V'] \, |S(V')| \sum_{i=1}^{|S|} 1_{\{|S(V')|=i\}} \\ &= \sum_{i=1}^{|S|} i \sum_{V' \subseteq V} \Pr[V'] \, 1_{\{|S(V')|=i\}} \end{aligned}$$

$$\begin{aligned} &= \sum_{i=1}^{|S|} i \binom{|S|}{i} p^i (1-p)^{|S|-i} \\ &\quad \times \sum_{j=0}^{n-|S|} \binom{n-|S|}{j} p^j (1-p)^{n-|S|-j} \\ &= p|S| \end{aligned}$$

where in the last summation we count all the subgraphs $G[V']$ such that $|S(V')| = i$ and we add their probabilities; also, $\sum_{j=0}^{n-|S|} C_j^{n-|S|} p^j (1-p)^{n-|S|-j} = 1$.

The above proof for $E(G, S, \texttt{M1})$ can be generalized in order to compute every moment of any order for M1. For instance:

$$\begin{aligned} E\left[(G, S, \texttt{M1})^2\right] &= \sum_{V' \subseteq V} \Pr[V'] \, |S_1'|^2 \\ &= \sum_{i=1}^{|S|} i^2 \binom{|S|}{i} p^i (1-p)^{|S|-i} = p|S|(p|S| + 1 - p) \end{aligned}$$

and, consequently:

$$\mathrm{Var}(G, S, \texttt{M1}) = E\left[(G, S, \texttt{M1})^2\right] - (E(G, S, \texttt{M1}))^2 = |S|p(1-p)$$

So, for M1, the random variable representing the size $|S|$ of the *a priori* solution follows a binomial law with parameters $|S|$ and p.

2.2.2. *Approximating optimal solutions*

In this section we show how, even if one cannot compute the optimal *a priori* solution in polynomial time, one can compute a suboptimal solution, the value (expectation) of which is always greater than a factor times the value (expectation) of the optimal solution. For this, we will propose in what follows well-known (in the theory of polynomial approximation of **NP**-hard problems) polynomial algorithms computing "good" suboptimal solutions, and will show that, also in the probabilistic case, these algorithms work well.

Recall that as we have already seen in section 2.2, PROBABILISTIC MAX INDEPENDENT SET1 is equivalent to a weighted MAX INDEPENDENT SET-problem, where each vertex is weighted by the corresponding probability. Consequently, the following theorem holds immediately.

THEOREM 2.4.– *If there exists a polynomial time approximation algorithm* **A** *solving* MAX WEIGHTED INDEPENDENT SET *within approximation ratio* ρ, *then* **A** *polynomially solves* PROBABILISTIC MAX INDEPENDENT SET*1 within the same approximation ratio* ρ.

In [DEM 99], an algorithm is developed for MAX WEIGHTED INDEPENDENT SET achieving approximation ratio:

$$\min\left\{\frac{\log n}{3(\Delta(G)+1)\log\log n}, O\left(n^{-\frac{4}{5}}\right)\right\}$$

Using this algorithm in Theorem 2.4, one gets the following corollary.

COROLLARY 2.1.– PROBABILISTIC MAX INDEPENDENT SET1 can be approximated within approximation ratio:

$$\min\left\{\frac{\log n}{3(\Delta(G)+1)\log\log n}, O\left(n^{-\frac{4}{5}}\right)\right\}$$

The characterization of PROBABILISTIC MAX INDEPENDENT SET1 in terms of a weighted MAX INDEPENDENT SET-problem draws not only issues for finding reasonable *a priori* suboptimal solutions, but unfortunately limits the capacity of the problem to be "well-approximated" since, via this characterization, all the negative results applying to MAX INDEPENDENT SET are immediately transferred to PROBABILISTIC MAX INDEPENDENT SET1 also. So, PROBABILISTIC MAX INDEPENDENT SET1 is hard to approximate within $n^{\epsilon-1}$, for any $\epsilon > 0$ ([HÅS 99]).

2.2.3. *Dealing with bipartite graphs*

In this section we study the complexity of solving PROBABILISTIC MAX INDEPENDENT SET1 in bipartite graphs. We show that, in this case, the problem dealt is polynomial.

THEOREM 2.5.– PROBABILISTIC MAX INDEPENDENT SET1 *is polynomial in bipartite graphs.*

Proof. Consider a bipartite graph $B(V_1, V_2, E_B)$. Then, by Theorem 2.2:

$$E(G, S, \mathtt{M1}) = \sum_{v_i \in S} p_i$$

Therefore, the optimal *a priori* solution $\hat{S}$ is a maximum-weight independent set in B considering that its vertices are weighted by the corresponding presence probabilities. Determining an optimal MAX INDEPENDENT SET-solution in a bipartite graph is of polynomial complexity in both weighted ([BOU 84]) and unweighted ([HAR 69]) cases (in [BOU 84], the polynomiality of weighted MAX INDEPENDENT SET in a class of graphs including the bipartite ones is proved). In the former case, this can be done in $O(n^{1/2}|E_B|)$. Finally, note that the complexity of the computation of $E(B, S, \mathtt{M1})$ is $O(n)$. ∎

2.3. PROBABILISTIC MAX INDEPENDENT SET2 and 3

2.3.1. *Expressions for $E(G, S, \mathtt{M2})$ and $E(G, S, \mathtt{M3})$*

Let $A_i = \sum_{V' \subseteq V, |S \cap V'| = i} \Pr[V']|S'_2|$. Then, $E(G, S, \mathtt{M2})$ can be easily written as:

$$\begin{aligned} E(G, S, \mathtt{M2}) &= \sum_{V' \subseteq V} \Pr[V'] \, |S'_2| \\ &= \sum_{V' \subseteq V} \left(\sum_{i=0}^{|S|} 1_{\{|S \cap V'| = i\}} \right) \Pr[V'] \, |S'_2| \\ &= \sum_{i=0}^{|S|} \left(\sum_{\substack{V' \subseteq V \\ |S \cap V'| = i}} \Pr[V'] \, |S'_2| \right) = \sum_{i=0}^{|S|} A_i \end{aligned} \qquad [2.4]$$

Quantities A_i, $i = 1, \ldots, |S|$ ($A_0 = 0$) are very natural and interesting from both theoretical and practical points of view. For instance, formula for $E(G, S, \mathtt{M2})$ given by the expression above holds for every probability law; also, computing analytical expressions for A_i seems to be an interesting problem in combinatorial counting of graphs; moreover, thanks to the simple relation between $E(G, S, \mathtt{M2})$ and A_i, $i = 1, \ldots, |S|$, analytical expressions for the latter would produce explicit expressions for the former. Unfortunately, [2.4], even intuitive and smart, does not give any hint which would allow precise characterization of $\hat{S}$.

In Proposition 2.1 below, the proof of which is in section 2.8.1 quantities $A_{\alpha(G)}$ and $A_{\alpha(G)-1}$ are explicitly computed, for the case of identical vertex-probabilities. However, the explicit computation for A_is of lower index produces very long and non-intuitive expressions.

PROPOSITION 2.1.– *Let* $\Gamma'(v) = \Gamma(v) \setminus \{\Gamma(v) \cap \Gamma(S^* \setminus \{v\})\}$, $\ell_1 = |\{v \in S^* : \Gamma'(v) = \emptyset\}|$, $\ell'_1 = \alpha(G) - \ell_1$, *and* $p_i = p$, $\forall v_i \in V$. *Then:*

$$\begin{aligned} A_{\alpha(G)} &= \alpha(G)p^{\alpha(G)} \\ A_{\alpha(G)-1} &= p^{\alpha(G)-1}(1-p) \\ &\times \left(\alpha(G)\ell_1 - \ell_1 + \alpha(G)\ell'_1 - \sum_{\substack{v \in S^* \\ \Gamma'(v) \neq \emptyset}} (1-p)^{|\Gamma'(v)|} \right) \end{aligned}$$

2.3.2. *An upper bound for the complexity of* $E(G, S, \mathtt{M2})$

We shall now give an upper bound for the complexity of computing $E(G, S, \mathtt{M2})$. For this, we will analyze (as an intermediate step) strategy M3 introduced in section 2.1.

Let $G'(V', E') = G[V \setminus S]$, and let $V' = \{v_1, \dots, v_{n-|S|}\}$ be the list of vertices of G' sorted in increasing-degree order; let us denote by V_i the set of the i first vertices of V' and let $G'_i = G'[V_i]$ (of course, for G'_i the vertices of V_i are not sorted in increasing-degree order). Let us denote by S'_i the independent set found by M3 on (the present subinstance of) G'_i, and by s'_i its cardinality, $i = 1, \dots, n - |S|$. The expression for the functional associated with M3 is given in Theorem 2.6 just below (its proof can be found in section 2.8.2).

THEOREM 2.6.– *If* $E(s'_i)$ *denotes the expectation of* s'_i, $i = 1, \dots, n - |S|$, *then:*

$$\begin{aligned} E\left(s'_{n-|S|}\right) &= \sum_{i=1}^{n-|S|} p_i \Pr\left[v_i \notin \Gamma\left(S'_{i-1}\right)\right] \\ E(G, S, \mathtt{M3}) &= E(G, S, \mathtt{M1}) + E\left(s'_{n-|S|}\right) \end{aligned}$$

Denoting by $T(E(s'_{n-|S|}))$ *and* $T(E(G, S, \mathtt{M3}))$ *the computation times of* $E(s'_{n-|S|})$ *and* $E(G, S, \mathtt{M3})$, *respectively:*

$$\begin{aligned} T\left(E\left(s'_{n-|S|}\right)\right) &= O\left(2^{n-|S|}\right) \\ T(E(G, S, \mathtt{M3})) &= O\left(2^{n-|S|}\right) \end{aligned}$$

Strategy M3 is, as has been already noted, a simplified version of algorithm M2. Moreover, there exist graphs where the two algorithms give the same results by performing identical choices and deletions of vertices (for instance, consider a graph on n isolated vertices). Consequently, computation time of M3 is a (worst-case) lower bound for the one of M2 and the following theorem holds.

THEOREM 2.7.– *Let $T(E(G, S, \texttt{M2}))$ be the computational time of $E(G, S, \texttt{M2})$. Then, $T(E(G, S, \texttt{M2})) = \Omega(T(E(G, S, \texttt{M3})))$.*

The result of Theorem 2.6 simply gives an upper bound on the complexity of computing $E(G, S, \texttt{M3})$ and does not prove that $E(G, S, \texttt{M3})$ is not computable in polynomial time (if this was true, it would be a very interesting result since, in this case, PROBABILISTIC MAX INDEPENDENT SET2 and PROBABILISTIC MAX INDEPENDENT SET3 would not belong to **NPO**; see section B.2 of Appendix B). In fact, the result of Theorem 2.6 is based upon a particular recursion-formula and a particular way for computing it. In any case, one can easily prove that PROBABILISTIC MAX INDEPENDENT SET2 is intractable (following the notation in the appendix of [GAR 79], PROBABILISTIC MAX INDEPENDENT SET2 is a kind of starry problem).

Indeed, if one can polynomially determine an optimal *a priori* solution $\hat{S}$ for PROBABILISTIC MAX INDEPENDENT SET2, then one can simply consider an instance of MAX INDEPENDENT SET as a PROBABILISTIC MAX INDEPENDENT SET2-instance with $p_i = 1$, $\forall v_i \in V$. It is easy to see that in this case, $\hat{S} = S^*$ and the following theorem immediately holds.

THEOREM 2.8.– *Unless **P=NP**,* PROBABILISTIC MAX INDEPENDENT SET2 *is computationally intractable.*

2.3.3. *Bounds for $E(G, S, \texttt{M2})$*

For lack of characterizing the complexity of computing $E(G, S, \texttt{M2})$, we build in this section upper and lower bounds for it. They are given in Theorem 2.9.

THEOREM 2.9.– *Let Δ be the maximum degree of $G[\tilde{V}']$. Then, on the hypothesis of distinct vertex-probabilities:*

$$\begin{aligned} E(G, S, \texttt{M2}) &\geqslant \sum_{v_i \in S} p_i + \sum_{V' \subseteq V} \prod_{v_i \in V'} p_i \prod_{v_i \notin V'} (1 - p_i) . \frac{|\tilde{V}'|}{\Delta+1} \\ E(G, S, \texttt{M2}) &\leqslant \sum_{v_i \in S} p_i + \sum_{V' \subseteq V} \prod_{v_i \in V'} p_i \prod_{v_i \notin V'} (1 - p_i) \left|\tilde{V}'\right| \end{aligned} \qquad [2.5]$$

while, on the hypothesis of identical vertex-probabilities:

$$\begin{aligned} E(G,S,\mathtt{M2}) &\geqslant p|S| + \sum_{V'\subseteq V} p^{|V'|}(1-p)^{n-|V'|}\frac{|\tilde{V}'|}{\Delta+1} \\ E(G,S,\mathtt{M2}) &\leqslant p|S| + \sum_{V'\subseteq V} p^{|V'|}(1-p)^{n-|V'|}\left|\tilde{V}'\right| \end{aligned} \quad [2.6]$$

Always under the latter hypothesis:

$$\frac{pn}{\Delta(G)+1} \leqslant E(G,S,\mathtt{M2}) \leqslant \frac{n(\Delta(G)+p)}{\Delta(G)+1}$$

Proof. Observe first that the following expression holds for $|S(V')|$:

$$\begin{aligned} |S(V')| \leqslant |S_2'| &\leqslant |S(V')| + |V' \setminus S(V') \cup (\Gamma(S(V')) \cap V')| \\ &= |S(V')| + \left|\tilde{V}'\right| \end{aligned} \quad [2.7]$$

Consequently, using [2.7], the following holds:

$$\begin{aligned} E(G,S,\mathtt{M2}) &\leqslant \sum_{V'\subseteq V} \Pr[V']\left(|S(V')| + \left|\tilde{V}'\right|\right) \\ &\leqslant E(G,S,\mathtt{M1}) + \sum_{V'\subseteq V} \Pr[V']\left|\tilde{V}'\right| \\ &= \sum_{v_i\in S} p_i + \sum_{V'\subseteq V} \Pr[V']\left|\tilde{V}'\right| \end{aligned} \quad [2.8]$$

On the other hand, $\Pr[V'] = \prod_{v_i\in V'} p_i \prod_{v_i\notin V'}(1-p_i)$. The upper bound results from the combination of [2.8] and the one for $\Pr[V']$.

We now prove the lower bound for $E(G,S,\mathtt{M2})$. Denote by $\tilde{S}_2$ the independent set computed by M2 when applied to $G[\tilde{V}']$ and by $|\tilde{S}_2|$ its cardinality. We then have:

$$\begin{aligned} E(G,S,\mathtt{M2}) &= \sum_{v_i\in S} p_i + \sum_{V'\subseteq V} \Pr[V']\left|\tilde{S}_2\right| \\ &= \sum_{v_i\in S} p_i + \sum_{V'\subseteq V} \prod_{v_i\in V'} p_i \prod_{v_i\notin V'} (1-p_i)\left|\tilde{S}_2\right| \end{aligned} \quad [2.9]$$

For $|\tilde{S}_2|$, since the greedy algorithm implied by M2 provides a maximal independent set, the following holds ([BER 73]): $|\tilde{S}_2| \geqslant |\tilde{V}'|/(\tilde{\Delta}+1)$. By substituting the expression for $|\tilde{S}_2|$ in [2.9], we obtain the lower bound claimed.

In the case where all the vertices have the same presence probability p, the following holds:

$$\sum_{v_i \in S} p_i \quad = \quad p|S|$$

$$\Pr\left[V'\right] = p^{|V'|}(1-p)^{n-|V'|}$$

and [2.6] follows immediately.

In order to obtain bounds implied by the last expression of the theorem (always assuming identical occurrence probabilities), we use inequality $|S| \geqslant n/(\Delta(G)+1)$. Moreover, $\sum_{V' \subseteq V} p^{|V'|}(1-p)^{n-|V'|} = 1$ (so, $\sum_{V' \subseteq V} p^{|V'|}(1-p)^{n-|V'|} \geqslant 0$), and:

$$\begin{aligned} \left|\tilde{V'}\right| &= |V' \setminus \{S(V') \cup \Gamma(S(V'))\}| \leqslant |V' \setminus S(V')| \\ &= |V' \setminus (S \cap V')| \leqslant |V \setminus (S \cap V)| \ = \ |V \setminus S| \ = \ n - |S| \end{aligned}$$

So, $\sum_{V' \subseteq V} p^{|V'|}(1-p)^{n-|V'|}|\tilde{V}'| \leqslant (n-|S|)\sum_{V' \subseteq V} p^{|V'|}(1-p)^{n-|V'|} = n-|S|$ and combining the above inequalities, we obtain the claimed bounds. ■

2.3.4. *Approximating optimal solutions*

2.3.4.1. *Using* $\mathbf{argmax}\{\sum_{v_i \in S} p_i\}$ *as an a priori solution*

Suppose that $\bar{S} = \text{argmax}\{\sum_{v_i \in S} p_i \ : \ S$ independent set of $G\}$ is used as an *a priori* solution for PROBABILISTIC MAX INDEPENDENT SET2 (recall that $\bar{S}$ *cannot be computed in polynomial time*). Then, the following holds.

THEOREM 2.10.– *Approximation of* $\hat{S}$ *by the solution:*

$$\bar{S} = \text{argmax}\left\{\sum_{v_i \in S} p_i \ : \ S \textit{ independent set of } G\right\}$$

guarantees approximation ratio bounded below by:

$$\max\left\{\frac{p_{\min}}{1+p_{\max}}, \frac{1}{\Delta(G)+1}\right\}$$

Proof. By [2.9] (proof of Theorem 2.9), and since $|\tilde{S}_2| \leqslant \alpha(G)$, we get:

$$\begin{aligned}\sum_{v_i \in S} p_i &\leqslant E(G,S,\mathtt{M2}) = \sum_{v_i \in S} p_i + \sum_{V' \subseteq V} \Pr[V'] \left|\tilde{S}_2\right| \\ &\leqslant \alpha(G)\left(p_{\max} + \sum_{V' \subseteq V} \Pr[V']\right) = \alpha(G)\left(1 + p_{\max}\right) \end{aligned} \quad [2.10]$$

Since $\bar{S} = \operatorname{argmax}\{\sum_{v_i \in S} p_i : S$ independent set of $G\}$, then:

$$E\left(G, \bar{S}, \mathtt{M2}\right) \geqslant \sum_{v_i \in \bar{S}} p_i \geqslant \sum_{v_i \in S^*} p_i \geqslant p_{\min}\alpha(G) \quad [2.11]$$

Note now that [2.10] holds also for $E(G, \hat{S}, \mathtt{M2})$; consequently, combining [2.10] and [2.11] we obtain:

$$\frac{E\left(G, \bar{S}, \mathtt{M2}\right)}{E\left(G, \hat{S}, \mathtt{M2}\right)} \geqslant \frac{p_{\min}\alpha(G)}{\left(1 + p_{\max}\right)\alpha(G)} = \frac{p_{\min}}{1 + p_{\max}} \quad [2.12]$$

On the other hand, note that from [2.2]:

$$E\left(G, \hat{S}, \mathtt{M2}\right) \leqslant \sum_{v_i \in V} p_i \quad [2.13]$$

and from the left-hand side of [2.5]:

$$E\left(G, \bar{S}, \mathtt{M2}\right) \geqslant \sum_{v_i \in \bar{S}} p_i \quad [2.14]$$

A combination of [2.13] and [2.14] gives:

$$\frac{E\left(G, \bar{S}, \mathtt{M2}\right)}{E\left(G, \hat{S}, \mathtt{M2}\right)} \geqslant \frac{\sum_{v_i \in \bar{S}} p_i}{\sum_{v_i \in V} p_i} \geqslant \frac{1}{\Delta(G) + 1} \quad [2.15]$$

where the last inequality (observe that $\bar{S}$ is maximal) is the weighted version of Turàn's Theorem ([TUR 41]).

Expressions [2.12] and [2.15] conclude the theorem. ∎

2.3.4.2. *Using approximations of* MAX INDEPENDENT SET

The set $\bar{S}$ considered in the previous section cannot be computed in polynomial time. Instead, suppose that one uses a polynomial time approximation algorithm A (achieving approximation ratio ρ) for (unweighted) MAX INDEPENDENT SET in order to compute a solution S' (obviously, we can suppose that S' is maximal) on G where vertex-probabilities are omitted. Then, [2.11] in the proof of Theorem 2.10 becomes

$$E\left(G, S', \mathtt{M2}\right) \geqslant p_{\min}\left|S'\right| \geqslant p_{\min}\rho\alpha(G)$$

and using the same arguments as in Theorem 2.10, the following theorem can be proved.

THEOREM 2.11.– *If there exists a polynomial time approximation algorithm* A *solving* MAX INDEPENDENT SET *within approximation ratio ρ, then algorithm* A *polynomially solves* PROBABILISTIC MAX INDEPENDENT SET2 *within approximation ratio:*

$$\max\left\{\frac{1}{\Delta(G)+1}, \left(\frac{p_{\min}}{1+p_{\max}}\right)\rho\right\}$$

If A *is the algorithm of [DEM 99], then:*

– *in the case of fixed vertex-probabilities,* PROBABILISTIC MAX INDEPENDENT SET2 *can be approximately solved in polynomial time within ratio:*

$$\min\left\{O\left(\frac{\log n}{3\left(\Delta(G)+1\right)\log\log n}\right), O\left(n^{-4/5}\right)\right\}$$

– *in the case where probabilities depend on n,* PROBABILISTIC MAX INDEPENDENT SET2 *is polynomially approximable within ratio:*

$$\max\left\{\frac{1}{\Delta(G)+1}, \left(\frac{p_{\min}}{1+p_{\max}}\right)O\left(\min\left\{\frac{\log n}{3\left(\Delta(G)+1\right)\log\log n}, n^{-\frac{4}{5}}\right\}\right)\right\}$$

2.3.5. *Dealing with bipartite graphs*

In this section we focus ourselves on bipartite graphs and study complexity of PROBABILISTIC MAX INDEPENDENT SET2 in this class of instances.

THEOREM 2.12.– *Consider a bipartite graph $B(V_1, V_2, E_B)$. Then:*

$$E\left(B, V_1, \mathtt{M2}\right) = \sum_{v_i \in V_1} p_i + \sum_{v_i \in V_2} p_i \prod_{v_j \in \Gamma(v_i)} \left(1 - p_j\right)$$

and can be computed in polynomial time. Consequently, in bipartite graphs, whenever color-class V_1 (or V_2) is considered as an a priori solution, PROBABILISTIC MAX INDEPENDENT SET2 *belongs to* ***NPO****.*

Proof. Let us first note the following:

– if all the vertices of V_1 are absent, then the solution provided by M2 is exactly the present vertices of the color-class V_2;

– if all the vertices of V_1 are present, then, despite the state of the set V_2, the solution of the present subinstance of B is exactly the color-class V_1;

– in the case where a part of the vertices of V_1 is present, the final solution for $B[V']$ will eventually include some vertices of V_2.

Applying the result of Theorem 2.1, we get:

$$E\left(B, V_1, \mathtt{M2}\right) = \sum_{v_i \in V_1} p_i + \sum_{v_i \in V_2} \Pr\left[X_i^{\mathtt{M2}, V_1} = 1\right] \qquad [2.16]$$

(recall that $\Pr[X_i^{\mathtt{M2}, S} = 1]$ represents the probability that vertex $v_i \notin S$ will be chosen when applying M2).

Note also that $\Pr[X_i^{\mathtt{M2}, V_1} = 1]$, $v_i \in V_2$, depends only on the present vertices of $\Gamma(v_i)$; consequently, it does not depend on the other elements of V_2. Henceforth, insertion of the elements of V_2 is performed independently of each other and $v_i \in V_2$ will be introduced in the solution for $B[V']$ only if $\Gamma(v_i) \cap V_1[V'] = \emptyset$. So, $\Pr[X_i^{\mathtt{M2}, V_1} = 1] = p_i \prod_{v_j \in \Gamma(v_i)} (1 - p_j)$.

Replacing this expression for $\Pr[X_i^{\mathtt{M2}, V_1} = 1]$ in [2.16], we obtain the result claimed for $E(B, V_1, \mathtt{M2})$. One can see that this expression implies the computation of $E(B, V_1, \mathtt{M2})$ in at most $O(n^2)$ steps. ∎

From the proof of Theorem 2.12, one can see how the particular structure of the bipartite graph intervenes in a significant way to simplify the expression for the functional and, consequently, its computation. Expression [2.16] holds thanks to the fact that the vertex set of B can be partitioned into two independent sets.

COROLLARY 2.2.– Suppose $\Pr[v_i] = p$, $v_i \in V_1 \cup V_2$, denote by n_1 and n_2 the sizes of V_1 and V_2, respectively, and suppose that $n_1 \geqslant n_2$. Then:

$$E\left(B, V_1, \mathtt{M2}\right) = pn_1 + p \sum_{v_i \in V_2} (1 - p)^{|\Gamma(v_i)|}$$

Naturally, $E(B, V_1, \mathtt{M2})$ is computable in $O(n^2)$.

From Corollary 2.2 we can obtain the following framing of $E(B, V_1, \mathtt{M2})$ for the case of identical vertex-probabilities:

$$\begin{aligned} E\left(B, V_1, \mathtt{M1}\right) + n_2 p(1-p)^{\Delta(B)} &\leqslant E\left(B, V_1, \mathtt{M2}\right) \\ &\leqslant E\left(B, V_1, \mathtt{M1}\right) + n_2 p(1-p)^{\delta(B)} \end{aligned}$$

So, for regular bipartite graphs (i.e., the ones where $\Delta(B) = \delta(B) = \Delta$):

$$E(B, V_1, \mathtt{M2}) = E(B, V_1, \mathtt{M1}) + n_2 p(1-p)^{\Delta}$$

Consider $\bar{S} = \text{argmax}\{\sum_{v_i \in S} p_i : S \text{ independent set of } B\}$ as an *a priori* solution for PROBABILISTIC MAX INDEPENDENT SET2. Then, considering vertex-probabilities as vertex-weights and using the result of [BOU 84], $\bar{S}$ can be computed in polynomial time. Furthermore:

$$E\left(B, \bar{S}, \mathtt{M2}\right) \geqslant \sum_{v_i \in \bar{S}} p_i \geqslant \frac{\sum_{v_i \in V} p_i}{2} \quad [2.17]$$

Using [2.17] together with [2.13], approximation ratio 1/2 is immediately yielded.

PROPOSITION 2.2.– $\bar{S} = \text{argmax}\{\sum_{v_i \in S} p_i : S$ *independent set of* $B\}$ *is a polynomial time approximation of* $\hat{S}$ *achieving approximation ratio 1/2 for* PROBABILISTIC MAX INDEPENDENT SET2 *in bipartite graphs.*

Obviously, from Theorem 2.12 and the discussion just above, the same approximation ratio can be yielded if one uses $\text{argmax}\{\sum_{v_i \in V_1} p_i, \sum_{v_i \in V_2} p_i\}$ as an *a priori* solution.

2.4. PROBABILISTIC MAX INDEPENDENT SET4

2.4.1. *An expression for* $E(G, S, \mathtt{M4})$

Recall that strategy M4 starts from $S(V')$ and completes it with the isolated vertices of the graph $G[\tilde{V}']$. In Proposition 2.3 below (its proof is in section 2.8.3), we give a polynomially computable expression for the functional associated with M4.

PROPOSITION 2.3.– *Given a graph* $G(V, E)$*, an a priori independent set* S *and the modification strategy M4, then, setting* $\Gamma_S(v_i) = \Gamma(v_i) \cap S$ *and* $\Gamma_{V \setminus S}(v_i) = \Gamma(v_i) \cap (V \setminus S)$*:*

$$\begin{aligned} E(G, S, \mathtt{M4}) \;&=\; \sum_{v_i \in S} p_i + \sum_{v_i \in (V \setminus S)} p_i \prod_{v_j \in \Gamma_S(v_i)} (1 - p_j) \\ &\times \prod_{v_k \in \Gamma_{V \setminus S}(v_i)} \left(1 - p_k \prod_{v_l \in \Gamma_S(v_k)} (1 - p_l)\right) \end{aligned} \quad [2.18]$$

$E(G, S, \mathtt{M4})$ can be computed in polynomial time. If $p_i = p$, for all $v_i \in V$, then:

$$E(G, S, \mathtt{M4}) = p|S| + \sum_{v_i \in (V \setminus S)} p(1-p)^{|\Gamma_S(v_i)|} \times \prod_{v_k \in \Gamma_{V \setminus S}(v_i)} \left(1 - p(1-p)^{|\Gamma_S(v_k)|}\right) \quad [2.19]$$

2.4.2. *Using S^* or $\mathrm{argmax}\{\sum_{v_i \in S} p_i\}$ as an a priori solution*

Expression [2.18], although polynomial, does not allow precise characterization of the optimal *a priori* solution $\hat{S}$ associated with M4. In Theorem 2.13 below (the proof of which is given in section 2.8.4), we restrict ourselves to the case of identical vertex probabilities and suppose that S^*, a maximum-size independent set of G, is used as an *a priori* solution[1]. Our objective is to estimate the ratio $E(G, S^*, \mathtt{M4})/E(G, \hat{S}, \mathtt{M4})$.

THEOREM 2.13.– *Under identical vertex-probabilities:*

$$\frac{E\left(G, S^*, \mathtt{M4}\right)}{E\left(G, \hat{S}, \mathtt{M4}\right)} \geqslant \frac{\alpha(G)}{n}\left(1 - (1-p)^{\Delta(G)}\right) + (1-p)^{\Delta(G)} \geqslant \frac{\alpha(G)}{n}$$

The ratio $\alpha(G)/n$ is always bounded below by $1/(\Delta(G)+1)$.

For instance, if G is cubic, i.e., $\Delta(G) = 3$, then $\alpha(G) \geqslant n/4$ and:

$$\frac{E\left(G, S^*, \mathtt{M4}\right)}{E\left(G, \hat{S}, \mathtt{M4}\right)} \geqslant \frac{1}{4}\left(1 - (1-p)^3\right) + (1-p)^3 \geqslant \frac{1}{4}$$

If in addition $p = 1/2$, then $E(G, S^*, \mathtt{M4})/E(G, \hat{S}, \mathtt{M4}) \geqslant 11/32$.

The result of Theorem 2.13 can be easily extended to the case where vertex-probabilities are distinct and $\bar{S} = \mathrm{argmax}\{\sum_{v_i \in S} p_i : S \text{ independent set of } G\}$ is used as an *a priori* solution. Moreover, without loss of generality, one can suppose that $\bar{S}$ is maximal. Then, one easily gets from [2.18]:

$$E\left(G, \hat{S}, \mathtt{M4}\right) \leqslant \sum_{v_i \in \hat{S}} p_i + \sum_{v_i \in \left(V \setminus \hat{S}\right)} p_i = \sum_{v_i \in V} p_i \quad [2.20]$$

$$E\left(G, \bar{S}, \mathtt{M4}\right) \geqslant \sum_{v_i \in \bar{S}} p_i \quad [2.21]$$

1. Recall once more that such an independent set cannot be computed in polynomial time.

Combining the above expressions, we obtain:

$$\frac{E\left(G, \bar{S}, \mathtt{M4}\right)}{E\left(G, \hat{S}, \mathtt{M4}\right)} \geqslant \frac{\sum_{v_i \in \bar{S}} p_i}{\sum_{v_i \in V} p_i} \geqslant \frac{1}{\Delta(G)+1}$$

where the last inequality is the weighted version of Turàn's theorem.

Finally, let us note that the same approximation ratio can be obtained if one treats vertex probabilities as weights and uses as an *a priori* solution the one computed by the greedy MAX INDEPENDENT SET-algorithm. In the weighted case, this algorithm iteratively chooses the vertex maximizing the ratio "vertex-weight over vertex-degree" and eliminates its neighbors. In this case, if S' is the independent set computed, we have ([PAS 97]):

$$E\left(G, S', \mathtt{M4}\right) \geqslant \sum_{v_i \in S'} p_i \geqslant \frac{\sum_{v_i \in V} p_i}{\Delta(G)+1}$$

and using [2.20], approximation ratio bounded below by $1/(\Delta(G)+1)$ is immediately concluded.

2.4.3. *Dealing with bipartite graphs*

The particular structure of a bipartite graph $B(V_1, V_2, E)$ does not allow refinement of the result of Proposition 2.3 in order to obtain a better characterization of the *a priori* solution maximizing $E(G, S, \mathtt{M4})$ and of the complexity of its computation.

However, if $\text{argmax}\{|V_1|, |V_2|\}$ is used as an *a priori* solution, then [2.18] can be simplified. Plainly, let us revisit it and suppose, without loss of generality, that $V_1 = \text{argmax}\{|V_1|, |V_2|\}$. So, we have $S = V_1$ and $V \setminus S = V_2$. Consequently, for $v_i \in V_2$, $\Gamma(v_i) \cap V_2 = \Gamma_{V \setminus S}(v_i) = \emptyset$ and the term $\prod_{v_k \in \Gamma_{V \setminus S}(v_i)}((1-p_k) + p_k(1 - \prod_{v_l \in \Gamma_S(v_k)}(1-p_l)))$ (the last product of [2.18]), computed on an empty set takes, by convention, value 1. So, dealing with bipartite graphs, [2.18] becomes:

$$E\left(B, V_1, \mathtt{M4}\right) = \sum_{v_i \in V_1} p_i + \sum_{v_i \in V_2} p_i \prod_{v_j \in \Gamma_{V_1}(v_i)} (1-p_j) \qquad [2.22]$$

In what follows, we will prove that if we use the color-class:

$$\text{argmax}\left\{\sum_{v_i \in V_1} p_i, \sum_{v_i \in V_2} p_i\right\}$$

as *a priori* solution for PROBABILISTIC MAX INDEPENDENT SET4, then it is solved within approximation ratio 1/2.

Indeed, let $V_1 = \text{argmax}\{\sum_{v_i \in V_1} p_i, \sum_{v_i \in V_2} p_i\}$. Then, by [2.22], we get:

$$E(B, V_1, \mathtt{M4}) \geqslant \sum_{v_i \in V_1} p_i \geqslant \frac{\sum_{v_i \in V_1 \cup V_2} p_i}{2} \quad [2.23]$$

Combining [2.23] with [2.20], we obtain:

$$\frac{E(B, V_1, \mathtt{M4})}{E\left(B, \hat{S}, \mathtt{M4}\right)} \geqslant \frac{\frac{\sum_{v_i \in V_1 \cup V_2} p_i}{2}}{\sum_{v_i \in V_1 \cup V_2} p_i} = \frac{1}{2}$$

The same worst case approximation ratio is also achieved if one sees probabilities as weights and considers the maximum-weight independent set (of total weight at least equal to $\sum_{v_i \in V_1} p_i$; this set can be found in polynomial time ([BOU 84])) as an *a priori* solution and the following theorem concludes the discussion above.

THEOREM 2.14.– *Given a bipartite graph B, the vertex-set:*

$$\text{argmax}\left\{\sum_{v_i \in V_1} p_i, \sum_{v_i \in V_2} p_i\right\}$$

as well as any maximum independent set[2] *of B are polynomial approximations of* PROBABILISTIC MAX INDEPENDENT SET*4 achieving approximation ratio bounded below by 1/2.*

Note finally that for the case of identical vertex-probabilities, the result of Theorem 2.14 could be obtained by direct combination of expressions [2.20] and [2.21].

2.5. PROBABILISTIC MAX INDEPENDENT SET5

2.5.1. *In general graphs*

Recall that strategy M5 considers the restriction $C(V')$ of an *a priori* vertex cover in the present subgraph $G[V']$ of G, it removes the isolated vertices (if any) from $C(V')$, and it finally takes the complement, with respect to V', of the resulting set.

2. Recall that a maximum independent set can be computed in polynomial time in a bipartite graph.

THEOREM 2.15.– *Given a graph $G(V, E)$, an a priori independent set S and the modification strategy* M5*, then:*

$$E(G, S, \texttt{M5}) = \sum_{v_i \in S} p_i + \sum_{v_i \in (V \setminus S)} p_i \prod_{v_j \in \Gamma(v_i)} (1 - p_j) \qquad [2.24]$$

$E(G, S, \texttt{M5})$ *is computable in polynomial time.*

Proof. Steps 1 to 3 of modification strategy M5 in section 2.1 constitute a modification strategy, denoted by M in what follows, for probabilistic vertex cover problem. For an *a priori* vertex cover C, the functional associated with M is (see Chapter 3 for a detailed computation):

$$E(G, C, \texttt{M}) = \sum_{v_i \in C} p_i - \sum_{v_i \in C} p_i \prod_{v_j \in \Gamma(v_i)} (1 - p_j) \qquad [2.25]$$

Using [2.25], we have the following for $E(G, S, \texttt{M5})$:

$$\begin{aligned} E(G, S, \texttt{M5}) &= \sum_{V' \subseteq V} \Pr[V'] \left|S'_5\right| = \sum_{V' \subseteq V} \Pr[V'] \left(|V'| - |C(V')|\right) \\ &= \sum_{V' \subseteq V} \Pr[V'] |V'| - \sum_{V' \subseteq V} \Pr[V'] |C(V')| \\ &= \left(\sum_{V' \subseteq V} \Pr[V'] \sum_{v_i \in V} 1_{\{v_i \in V'\}} \right) - E(G, V \setminus S, \texttt{M}) \\ &= \left(\sum_{v_i \in V} \sum_{V' \subseteq V} \Pr[V'] 1_{\{v_i \in V'\}} \right) - E(G, V \setminus S, \texttt{M}) \\ &= \sum_{v_i \in V} p_i - \sum_{v_i \in (V \setminus S)} p_i + \sum_{v_i \in (V \setminus S)} p_i \prod_{v_j \in \Gamma(v_i)} (1 - p_j) \\ &= \sum_{v_i \in S} p_i + \sum_{v_i \in (V \setminus S)} p_i \prod_{v_j \in \Gamma(v_i)} (1 - p_j) \end{aligned}$$

and the proof of the theorem is complete. ∎

Expression [2.24] can be rewritten as:

$$E(G,S,\mathtt{M5}) = \sum_{v_i \in V} p_i - \sum_{v_i \in (V \setminus S)} p_i + \sum_{v_i \in (V \setminus S)} p_i \prod_{v_j \in \Gamma(v_i)} (1-p_j)$$
$$= \sum_{v_i \in V} p_i - \sum_{v_i \in (V \setminus S)} p_i \left(1 - \prod_{v_j \in \Gamma(v_i)} (1-p_j)\right)$$

Since, for a graph G, quantity $\sum_{v_i \in V} p_i$ is constant, maximization of $E(G,S,\mathtt{M5})$ becomes equivalent to the minimization of $\sum_{v_i \in (V \setminus S)} p_i(1 - \prod_{v_j \in \Gamma(v_i)}(1-p_j))$. But S being a maximal independent set, $V \setminus S$ is a minimal vertex covering of G, and in order to find the vertex covering C minimizing the quantity:

$$\sum_{v_i \in C} p_i \left(1 - \prod_{v_j \in \Gamma(v_i)} (1-p_j)\right)$$

one has simply to consider each vertex $v_i \in V$ as weighted by the weight:

$$w_i = p_i \left(1 - \prod_{v_j \in \Gamma(v_i)} (1-p_j)\right)$$

and to search for a minimum-weight vertex cover. Consequently the following theorem characterizes the *a priori* solution maximizing $E(G,S,\mathtt{M5})$.

THEOREM 2.16.– *The a priori solution $\hat{S}$ maximizing $E(G,S,\mathtt{M5})$ is the complement, with respect to V, of a minimum-weight vertex cover of G where every vertex v_i is weighted by a weight $p_i(1 - \prod_{v_j \in \Gamma(v_i)}(1-p_j))$. Consequently,* PROBABILISTIC MAX INDEPENDENT SET5 *is **NP**-hard.*

In other words, Theorem 2.16 establishes that, as in the case of PROBABILISTIC MAX INDEPENDENT SET1, PROBABILISTIC MAX INDEPENDENT SET5 is equivalent to a MAX WEIGHTED INDEPENDENT SET. Since weights do not intervene in the ratio obtained in [DEM 99], Corollary 2.1 holds also for PROBABILISTIC MAX INDEPENDENT SET5.

2.5.2. *In bipartite graphs*

Since maximum-weight independent is polynomial in bipartite graphs ([BOU 84]), so is minimum-weight vertex covering. So the following theorem immediately holds.

THEOREM 2.17.– *The a priori solution $\hat{S}$ maximizing $E(B, S, \mathtt{M5})$ is the complement, with respect to $V_1 \cup V_2$, of a minimum-weight vertex cover of B where every vertex v_i is weighted by a weight $p_i(1 - \prod_{v_j \in \Gamma(v_i)}(1 - p_j))$. Consequently,* PROBABILISTIC MAX INDEPENDENT SET5 *is polynomial for bipartite graphs.*

2.6. Summary of the results

Dealing with the quality of the solutions obtained, we have the following relation for a fixed *a priori* solution S:

$$E\left(G, S, \mathtt{M1}\right) \leqslant E(G, S, \mathtt{M5}) \leqslant E(G, S, \mathtt{M4}) \leqslant E\left(G, S, \mathtt{M2}\right) \qquad [2.26]$$

First inequality in [2.26] is obvious and follows from Theorem 2.2 and [2.24] in Theorem 2.15. In order to prove the second inequality, observe that (using [2.18]):

$$p_k \left(1 - \prod_{v_l \in \Gamma_S(v_k)} (1 - p_l)\right) \geq 0$$

and consequently:

$$\begin{aligned} E(G, S, \mathtt{M4}) &\geqslant \sum_{v_i \in S} p_i + \sum_{v_i \in (V \setminus S)} p_i \times \prod_{v_j \in \Gamma_S(v_i)} (1 - p_j) \\ &\quad \times \prod_{v_k \in \Gamma_{V \setminus S}(v_i)} (1 - p_k) \\ &\geqslant \sum_{v_i \in S} p_i + \sum_{v_i \in (V \setminus S)} p_i \times \prod_{v_j \in \left(\Gamma_S(v_i) \cup \Gamma_{V \setminus S}(v_i)\right)} (1 - p_j) \\ &= \sum_{v_i \in S} p_i + \sum_{v_i \in (V \setminus S)} p_i \times \prod_{v_j \in \Gamma(v_i)} (1 - p_j) \\ &= E(G, S, \mathtt{M5}) \end{aligned}$$

Last inequality in [2.26] is due to the fact that for every subgraph $G[V']$, the cardinality of the solution computed applying M2 will be greater than the one of the solution computed applying M4 since GREEDY (called by M2) will add in the solution produced by applying M4 the isolated vertices of $G[\tilde{V}']$ (at least).

	Variant 1	Variant 2	Variant 4	Variant 5
$T(E)$	$O(n)$	$O\left(2^{n-\lvert S\rvert}\right)$	$O\left(n^2\right)$	$O\left(n^2\right)$
$\hat{S}$	MAX WIS(G_p^w)	hard	?	MAX WIS$(G_{p'}^w)$
Complexity	**NP**-hard	?	?	**NP**-hard

Table 2.1. *Complexities of computing functionals and characterizations and complexities of computing a priori solutions for several variants of probabilistic independent set in general graphs*

	Variant 1	Variant 2	Variant 4	Variant 5
$T(E)$	$O(n)$	?	$O\left(n^2\right)$	$O\left(n^2\right)$
$\hat{S}$	MAX WIS(B_p^w)	?	?	MAX WIS$(B_{p'}^w)$
Complexity	**P**	?	?	**P**

Table 2.2. *Complexities of computing functionals and characterizations and complexities of computing a priori solutions for several variants of probabilistic independent set in bipartite graphs*

Tables 2.1 and 2.2 summarize the main results of this chapter about the complexities of computing the functionals and the ones of computing the *a priori* solutions

	S	Approximation ratio
Variant 1	The one computed in [DEM 99]	$\mathbf{r}$
Variant 2	The one computed in [DEM 99]	$\max\left\{\left(\frac{p_{\min}}{1+p_{\max}}\right)\mathbf{r}, \frac{1}{\Delta(G)+1}\right\}$
	[DEM 99], probabilities independent of n	$\mathbf{r}$
	$\operatorname{argmax}\{\sum_{v_i \in S} p_i\}$	$\max\left\{\frac{p_{\min}}{1+p_{\max}}, \frac{1}{\Delta(G)+1}\right\}$
	S^*, identical probabilities	$\frac{\alpha(G)}{n}$
Variant 4	$\operatorname{argmax}\{\sum_{v_i \in S} p_i\}$	$\frac{1}{\Delta(G)+1}$
	The output of the greedy algorithm	$\frac{1}{\Delta(G)+1}$
Variant 5	The one computed in [DEM 99]	$\mathbf{r}$

Table 2.3. *Approximating a priori solutions in general graphs;* $\mathbf{r} = \min\{O(\log n/(3(\Delta(G)+1)\log\log n)), O(n^{-4/5})\}$

	S	Approximation ratio
Variant 2	$\text{argmax}\{\sum_{v_i \in S} p_i\}$	$\frac{1}{2}$
Variant 2	$\text{argmax}\{\sum_{v_i \in V_1} p_i, \sum_{v_i \in V_2} p_i\}$	$\frac{1}{2}$
Variant 4	$\text{argmax}\{\sum_{v_i \in S} p_i\}$	$\frac{1}{2}$
Variant 4	$\text{argmax}\{\sum_{v_i \in V_1} p_i, \sum_{v_i \in V_2} p_i\}$	$\frac{1}{2}$

Table 2.4. *Approximating a priori solutions in bipartite graphs*

maximizing them. In this tables, we denote by G_p^w (resp. B_p^w) a graph G (resp., a bipartite graph B) whose vertices are weighted by their corresponding probabilities, by $G_{p'}^w$ (resp., $B_{p'}^w$) a graph (resp., bipartite graph) whose vertex v_i is weighted by the quantity $p_i(1 - \prod_{v_j \in \Gamma(v_i)}(1 - p_j))$, $1 \leqslant i \leqslant n$, by $T(E)$ the time needed for the computation of the functional E and, for economy, by MAX WIS(G_p^w) (resp., MAX WIS($G_{p'}^w$)), the fact that the *a priori* solution maximizing the functional is a maximum-weight independent set in G_p^w (resp., $G_{p'}^w$). Finally, G and B denote general and bipartite graphs, respectively, and question marks denote open questions. Dealing with entry ($T(E)$,Variant 2) of Table 2.2, recall that the complexity of computing the functional for PROBABILISTIC MAX INDEPENDENT SET2 is polynomial if the set:

$$\text{argmax}\{|V_1|, |V_2|\}$$

is used as an *a priori* solution.

In Table 2.3, **r** stands for $\min\{O(\log n/(3(\Delta(G) + 1)\log\log n)), O(n^{-4/5})\}$. Here, as well as in Table 2.4, a summary of the main approximation results in general and in biparite graphs, respectively, is presented. Let us note that the approximation ratio in the line for Variant 5 of Table 2.3 is directly obtained with arguments exactly analogous to the ones of Corollary 2.1 in section 2.2.2, considering $p_i(1 - \prod_{v_j \in \Gamma(v_i)}(1 - p_j))$ as vertex-weight for v_i, $i = 1, \ldots, n$.

2.7. Methodological questions

As we have seen, PROBABILISTIC MAX INDEPENDENT SET1 and PROBABILISTIC MAX INDEPENDENT SET5 can be expressed as particular versions of MAX WEIGHTED INDEPENDENT SET. Then, one can ask her/himself if mathematical expectation is the best and most representative functional in order to represent the objective function of this problem. Let us take, for instance, PROBABILISTIC MAX INDEPENDENT SET1. Obviously, considering mathematical expectation $\max_S \sum_{v_i \in V} 1_{\{v_i \in S\}} p_i$ as the functional to be maximized is a fair and natural assumption. But, other functionals could be considered, depending on the nature and the context of the natural application modeled in terms of the problem dealt. For example, consider an application where vertices represent a gain in a game. A venturous player's behavior would consist of maximizing its chances, i.e., in finding the maximum cardinality independent set and then taking into account probabilities in order to re-adjust the solution computed. A more conservative player's behavior would be, on the other hand, to minimize the risks, i.e., to minimize the probability of having small gains, this behavior being, possibly, expressed by $\max_S \sum_{v_i \in V} 1_{\{v_i \in S: p_i \geqslant p_0\}}$, where p_0 is some threshold-probability, or yet by $\min_S \sum_{v_i \in V \setminus S} 1_{\{v_i \in V \setminus S: p_i \leqslant p_0\}}$. In these cases also, there exist many details to be discussed concerning the role played by the probabilities. In all, for PROBABILISTIC MAX INDEPENDENT SET1 (this remains true for all the other probabilistic independent set-variants dealt in this chapter) a solution including, for example, 10 vertices any of probability 0.9 is not the same as a solution including 90 vertices of probability 0.1, even if the mathematical expectations are identical for the two solutions. In fact, it is the context of the application (modeled in probabilistic terms) and the judicious choice of the algorithm that will determine which of the two solutions is the best for a given application.

Commonly, a combinatorial optimization problem models a concrete real-world application. In order to solve it, one has to determine a solution that satisfies a certain optimality criterion. In the framework of decision theory, one can interpret a feasible solution as a kind of decision. The choice of the optimality criterion determines the underlying probabilistic combinatorial optimization problem in the sense that different criteria induce distinct such problems. Several criteria can be provided; for example, the cost's expectation, the utility's expectation, the minimum gain, the maximum regret, etc. From a decisional point of view, the great difficulty is to choose which criterion is the most representative of the behavior and the preferences of a decision maker. When dealing with probabilistic combinatorial optimization graph-problems, any of the 2^n possible subgraphs of the initial graph represents a possible state of nature.

As we have seen in Chapter 1, PROBABILISTIC MAX INDEPENDENT SET can model a satellite shot planning problem. The decision here amounts to choosing, in a random universe, a certain number of shots to be realized. The optimality criterion we have used until now consists of maximizing the expected number of shots. We have seen above that, under some modification strategies, this criterion allows us to interpret vertex-probabilities in terms of vertex-weights and to reduce the probabilistic problem to a particular kind of weighted one. As we will see in the next chapters, for a certain number of problems this trend to reduce probabilities to weights is closely linked to the particular optimality criterion chosen. However, this particular point of view may be somewhat restrictive. We will sketch in this section how optimal solution evolves when other criteria, more "classical" in decision theory, are used. For simplicity, we will restrict ourselves to the modification strategies M1 and M2. Let us note that even if our discussion deals here with PROBABILISTIC MAX INDEPENDENT SET, its corollaries apply to any other of the problems studied in this book.

2.7.1. *Maximizing a criterion associated with gain*

In this case, one has first to define a notion of gain associated with a solution and a subgraph of the initial input-graph. Given a graph $G(V,E)$, a subset $V' \subseteq V$, an *a priori* solution S for G and a modification strategy $\mathtt{M}$, the gain $g(G[V'],S,M)$, associated with $G[V']$, S and $\mathtt{M}$, is determined by the value of the solution of $G[V']$ induced by the application of $\mathtt{M}$ to S. Then, we study two criteria commonly used in decision theory: the *minimum gain* and the *maximum gain*.

2.7.1.1. *The minimum gain criterion*

This criterion, associated with an *a priori* solution S, will be denoted by $f_g(S)$ and will be defined as:

$$f_g(S) = \min_{G[V']} \{g\left(G\left[V'\right], S, \mathtt{M}\right)\} = \min_{V' \subseteq V} \{g\left(G\left[V'\right] S, \mathtt{M}\right)\}$$

It expresses a certain conservatism of the decision maker since it amounts to evaluating a solution (decision) from a worst case performance point of view. With such criterion, one tends to consider the least risky of the possible decisions. An optimal solution S^* for this criterion is one verifying: $f_g(S^*) = \max_S\{f_g(S)\}$. This value does not represent the real gain associated with S^*. In order to determine it, one has to compute, for a given state V' of nature, the gain $g(G[V'], S^*, \mathtt{M})$.

2.7.1.1.1. Using modification strategy $\mathtt{M1}$

In this case, the gain considered is defined as:

$$g\left(G\left[V'\right], S, \mathtt{M1}\right) = |S \cap V'| \quad [2.27]$$

Hence, the minimum gain criterion associated with S is given by:

$$f_g(S) = \min_{V' \subseteq V} \{g\left(G\left[V'\right], S, \mathtt{M1}\right)\} = \min_{V' \subseteq V} \{|V' \cap S|\} = 0$$

Consequently, all the feasible *a priori* solutions are equivalent under this criterion, since they all have the same value. In other words, for a decision maker, all the solutions are identical because, for any of them, there exists at least a graph $G[V']$ for which the solution induced is empty. So, any feasible solution under the criterion at hand, is an optimal one.

2.7.1.1.2. Using modification strategy M2

Recall that, as we have seen in section 2.1, strategy M2 completes the solution computed by M1, in order to make it maximal for the inclusion. Using notations adopted previously, the gain associating with M2 can be expressed as:

$$g\left(G\left[V'\right], S, \mathtt{M2}\right) = |S \cap V'| + \left|S\left(\tilde{V}'\right)\right|$$

In this case, the criterion of the minimum gain becomes:

$$\begin{aligned} f_g(S) &= \min_{V' \subseteq V} \left\{g\left(G\left[V'\right], S, \mathtt{M2}\right)\right\} = \min_{V' \subseteq V} \left\{|S \cap V'| + \left|S\left(\tilde{V}'\right)\right|\right\} \\ &= \min_{V' \subseteq V} \left\{\left|(S \cap V') \cup S\left(\tilde{V}'\right)\right|\right\} \end{aligned} \qquad [2.28]$$

Here, if $V' = \emptyset$ is a feasible state of nature for the application modelled, then $f_g(S) = 0$. If not, then the optimal solution S^* verifies:

$$f_g\left(S^*\right) = \max_S \left\{\min_{V' \subseteq V} \left\{\left|(S \cap V') \cup S\left(\tilde{V}'\right)\right|\right\}\right\}$$

From an operational point of view, use of M2 assumes that one precisely knows set V', since one has also to know the subgraph $G[\tilde{V}']$.

What can be concluded from the discussion above is that minimum gain criterion seems to be quite senseless. In order to remedy this, one has then to make some more restrictive hypotheses on V'. For instance, we can assume that the feasible states of nature correspond to subgraphs whose orders are fractions of the order of the initial graph. But, in this case, one loses the hypothesis that vertex-probabilities are mutually independent.

2.7.1.2. *The maximum gain criterion*

This criterion, associated with an *a priori* solution S and denoted by $f^g(S)$, can be defined as:

$$f^g(S) = \max_{G[V']} \left\{g\left(G\left[V'\right], S, \mathtt{M}\right)\right\} = \max_{V' \subseteq V} \left\{g\left(G\left[V'\right], S, \mathtt{M}\right)\right\}$$

It corresponds to an optimistic behavior of the decision maker, since she/he evaluates a solution from a best case performance point of view. A solution S^* is optimal for this criterion, if it verifies $f^g(S^*) = \max_S\{f^g(S)\}$.

2.7.1.2.1. Using strategy M1

Using [2.27], the value of a solution S, under the criterion of maximum gain is given by:

$$f^g(S) = \max_{V' \subseteq V} \{g(G[V'], S, \texttt{M1})\} = \max_{V' \subseteq V} \{|S \cap V'|\} = |S|$$

For the optimal solution S^*, under the criterion at hand we have:

$$f^g(S^*) = \max_S \{f^g(S)\} = \max_S \{|S|\}$$

i.e., S^* is a maximum independent set of the initial graph G.

2.7.1.2.2. Using strategy M2

In an analogous way as for [2.28], the value of an *a priori* solution S for the maximum gain criterion is expressed by:

$$f^g(S) = \max_{V' \subseteq V} \{g(G[V'], S, \texttt{M2})\} = \max_{V' \subseteq V} \left\{\left|(S \cap V') \cup S\left(\tilde{V'}\right)\right|\right\} \quad [2.29]$$

PROPOSITION 2.4.– *The solution S^* maximizing $f^g(S)$ in [2.29] is a maximum independent set of the initial graph G.*

Proof. We first prove that, given a maximum independent set S^* of G and any subset $V' \subseteq V$:

$$\left|(S^* \cap V') \cup S^*\left(\tilde{V'}\right)\right| \leqslant |S^*| \quad [2.30]$$

Indeed, in the opposite case, there would exist a subgraph $G[V']$ of G such that $|(S^* \cap V') \cup S^*(\tilde{V'})| > |S^*|$, i.e., $G[V']$ should contain an independent set S_0 such that $|S_0| > |S^*|$, a contradiction since S^* is supposed to be maximum for G.

On the other hand, one can easily show that, for any $V' \subseteq V$, if $S^* \subseteq V'$, then:

$$\left|(S^* \cap V') \cup S^*\left(\tilde{V'}\right)\right| = |S^*| \quad [2.31]$$

Combination of [2.30] and [2.31] concludes that, for a maximum independent set S^* of G, $f^g(S^*) = \max_{V' \subseteq V}\{|(S \cap V') \cup S(\tilde{V'})|\} = |S^*|$.

It remains now to show that $f^g(S^*) = \max_S\{f^g(S)\}$. Let us notice that, for any S, $f^g(S^*) \geqslant f^g(S)$ if and only if $|S^*| \geqslant \max_{V' \subseteq V}\{|(S \cap V') \cup S(\tilde{V}')|\}$. This last inequality is always verified since, *a contrario*, there would exist a set $V_0 \subseteq V$ and an *a priori* solution S_0 for $G[V_0]$ such that $|S_0| > |S^*|$, a contradiction since S^* has been assumed to be maximum. The equivalence above concludes the proof. ∎

It should be noted that [2.30] is not systematically verified if S^* is not maximum. For instance, consider a bipartite graph $B(V_1, V_2, E)$, with $|V_2| > |V_1|$ and set $S = V_1$ (obviously, S is maximal but not maximum). If we consider $V' = V_2$, we have: $|(S \cap V') \cup S(\tilde{V}')| = |(V_1 \cap V_2) \cup V_2| = |V_2| > |V_1|$. Hence, for a non-maximum solution S, we do not necessarily have $f^g(S) = |S|$.

What is interesting to notice from the discussion in this section is that, in the opposite to what has been observed previously, when the functional was used as optimality criterion, the maximum gain criterion induces that both strategies M1 and M2 admit the same optimal *a priori* solutions.

2.7.2. *Minimizing a criterion associated with regret*

For any subgraph $G[V']$ induced by a subset V' of V, the best possible gain associated with it and with a modification strategy M is defined by:

$$g^*\left(G\left[V'\right], \mathtt{M}\right) = \max_S\left\{g\left(G\left[V'\right], S, \mathtt{M}\right)\right\} = \operatorname{opt}\left(G\left[V'\right]\right)$$

Then, the regret associated with a graph $G[V']$, an *a priori* solution S and a modification strategy M is given by:

$$r\left(G\left[V'\right], S, \mathtt{M}\right) = \operatorname{opt}\left(G\left[V'\right]\right) - g\left(G\left[V'\right], S, \mathtt{M}\right)$$

For any $G[V']$, this regret determines what one loses by choosing S as an *a priori* solution instead of the one for which its modification via M would result in $\operatorname{opt}(G[V'])$.

2.7.2.1. *The maximum regret criterion*

For an *a priori* solution S, this criterion is given by:

$$f^r(S) = \max_{V' \subseteq V}\left\{r\left(G\left[V'\right], S, \mathtt{M}\right)\right\}$$

Once more, $f^r(S)$ evaluates a solution S following a worst case point of view. The decision maker whose behavior fits this criterion estimates an action by placing her/himself in the most unfavorable configuration; one considers the maximum regret one risks getting by choosing S.

Maximum regret seems to be reasonable criterion when no complete information is available about how likely is any of the different states of nature, or when the decision has to be made only once. One of the major difficulties appearing when using maximum regret consists of determining, for any subgraph $G[V']$, the best possible solution. A solution S^* is considered optimal for the maximum regret criterion if it verifies:

$$f^r(S^*) = \min_S \{f^r(S)\} = \min_S \left\{ \max_{V' \subseteq V} \{g^*(G[V'], \mathtt{M}) - g(G[V'], S, \mathtt{M})\} \right\}$$

2.7.2.1.1. Using strategy $\mathtt{M1}$

Under modification strategy $\mathtt{M1}$, $g^*(G[V'], \mathtt{M1}) = \max_S\{|S \cap V'|\} = \text{opt}(G[V'])$. Indeed, we can always find a solution S such that, S restricted to $G[V']$ corresponds to an optimal solution for this subgraph. Note that this is not the same *a priori* solution S of G that always provides the optimal solution for any subgraph $G[V']$.

For an *a priori* solution S and a subgraph $G[V']$ of the initial graph G, the regret associated with $\mathtt{M1}$ is defined as: $r(G[V'], S, \mathtt{M1}) = \text{opt}(G[V']) - |S \cap V'|$. Consequently, the corresponding maximum regret is:

$$\begin{aligned} f^r(S) &= \max_{V' \subseteq V} \{r(G[V'], S, \mathtt{M1})\} \\ &= \max_{V' \subseteq V} \{\text{opt}(G[V']) - |S \cap V'|\} \end{aligned}$$

Let $S^* = \text{argmin}_S\{f^r(S)\}$. Then, $f^r(S^*)$ is defined by:

$$\begin{aligned} f^r(S^*) &= \min_S \{f^r(S)\} \\ &= \min_S \left\{ \max_{V' \subseteq V} \{\text{opt}(G[V']) - |S \cap V'|\} \right\} \end{aligned} \qquad [2.32]$$

2.7.2.1.2. Using strategy $\mathtt{M2}$

Obviously, with analogous arguments as in section 2.7.2.1.1, we have dealing with strategy $\mathtt{M2}$: $g^*(G[V'], \mathtt{M2}) = \max_S\{|S \cap V'| + |S(\tilde{V}')|\} = \text{opt}(G[V'])$. So, the regret associated with an *a priori* solution S, a subgraph $G[V']$ and with $\mathtt{M2}$ is defined by: $r(G[V'], S, \mathtt{M2}) = \text{opt}(G[V']) - (|S \cap V'| + |S(\tilde{V}')|)$ and, consequently, the maximum regret associated with S can be expressed as:

$$f^r(S) = \max_{V' \subseteq V} \left\{ \text{opt}(G[V']) - \left(|S \cap V'| + \left| S\left(\tilde{V}'\right) \right| \right) \right\}$$

Note that the *a priori* solution dealing with $\mathtt{M2}$ is "better" than the one dealing with $\mathtt{M1}$, since $r(G[V'], S, \mathtt{M2}) = r(G[V'], S, \mathtt{M1}) - |S(\tilde{V}')|$. This implies that regret under $\mathtt{M2}$ is less important than under $\mathtt{M1}$.

2.7.3. *Optimizing expectation*

The reader has certainly already noticed that in both of sections 2.7.1 and 2.7.2, the probabilistic nature of the problems dealt is not taken into account. Indeed, the use of criteria like the ones discussed there supposes that decision space is either not probabilistic, or that the randomness of this space is badly quantifiable or not quantifiable at all. If these are not the cases, using such criteria amounts to impoverishing the real models.

So, one has to build criteria taking into account as well as possible the probabilistic nature of the decision space. A natural criterion is then the mathematical expectation of the gain.

Let $g(G[V'], S, \mathtt{M})$ be the gain associated with a subgraph $G[V']$ of G induced by a set $V' \subseteq V$, an *a priori* solution S and a modification strategy $\mathtt{M}$. Then, the mathematical expectation of the gain, denoted by $E_g(S, \mathtt{M})$ is defined by:

$$E_g(S, \mathtt{M}) = \sum_{V' \subseteq V} g\left(G\left[V'\right], S, \mathtt{M}\right) \Pr\left[V'\right] \tag{2.33}$$

where $\Pr[V']$ denotes the occurrence-probability of the subgraph $G[V']$. Note that what has been done in this chapter moves, in fact, around [2.33], considering that the gain of a solution is its size.

Dealing with expectation of the regret, as this notion has been expressed and used in section 2.7.2, the following holds:

$$\begin{aligned} E_r(S, \mathtt{M}) &= \sum_{V' \subseteq V} r\left(G\left[V'\right], S, \mathtt{M}\right) \Pr\left[V'\right] \\ &= \sum_{V' \subseteq V} \operatorname{opt}\left(G\left[V'\right]\right) \Pr\left[V'\right] - \sum_{V' \subseteq V} g\left(G\left[V'\right], S, \mathtt{M}\right) \Pr\left[V'\right] \end{aligned} \tag{2.34}$$

Since the first term in [2.34] is independent from the *a priori* solution S under consideration, this expression becomes:

$$\min_S \left\{E_r(S, \mathtt{M})\right\} = \sum_{V' \subseteq V} \operatorname{opt}\left(G\left[V'\right]\right) \Pr\left[V'\right] - \max_S \left\{E_g(S, \mathtt{M})\right\} \tag{2.35}$$

2.8. Proofs of the results

2.8.1. *Proof of Proposition 2.1*

In order to determine $A_{\alpha(G)}$, we first need the following preliminary lemma.

LEMMA 2.1.– *If $S \subseteq V'$, then $|S'_2| = |S|$.*

Proof. By definition of M2, $|S'_2| = |S(V')| + |S(\tilde{V}')|$. But if $S \subseteq V'$, then $S(V') = S$ and this implies $S(V') \cup \Gamma(S(V')) = S \cup \Gamma(S)$. Moreover, the maximality of S implies that $S \cup \Gamma(S) = V$; and consequently, $S(\tilde{V}') = \emptyset$. So, $|S'_2| = |S(V')| = |S|$ and this completes the proof of Lemma 2.1. ∎

Determining $\boldsymbol{A_{\alpha(G)}}$

For $A_{\alpha(G)}$ we have $|S^* \cap V'| = \alpha(G)$ and, by Lemma 2.1:

$$m\left(G\left[V'\right], S^*\left(V', \mathtt{M2}\right)\right) = \alpha(G)$$

Therefore:

$$\begin{aligned}
A_{\alpha(G)} &= \alpha(G)\left(\sum_{\substack{V' \subseteq V \\ |S^* \cap V'| = \alpha(G)}} \Pr\left[V'\right]\right) \\
&= \alpha(G) p^{\alpha(G)} \sum_{i=0}^{n-\alpha(G)} \binom{n-\alpha(G)}{i} p^i (1-p)^{n-\alpha(G)-i} \\
&= \alpha(G) p^{\alpha(G)}
\end{aligned}$$

where, in the above expression, the term $p^{\alpha(G)}$ represents the fact that the $\alpha(G)$ vertices of S^* are all present in V' ($S^* \cap V' = S^*$) and the term $\sum_{i=0}^{n-\alpha(G)} C_i^{n-\alpha(G)} p^i (1-p)^{n-\alpha(G)-i}$ stands for all possible choices for the rest of the elements of V (as we have already seen, this term equals 1).

Determining $\boldsymbol{A_{\alpha(G)-1}}$

Consider an element v of S^* and note that since S^* is a maximum independent set, $\Gamma'(v)$ is either empty, or a clique on at least one vertex (if not, S^* could be augmented to $(S^* \setminus \{v\}) \cup \Gamma'(v)$). Consequently, considering that v is the element of S^* which is absent from V', and denoting by S_2^* the output of M2 when called with *a priori* solution S^*, we have:

– if $V' \cap \Gamma'(v) = \emptyset$, then the independent set $S^* \setminus \{v\}$ remains a maximal one for $G[V']$, so $|S_2^*| = \alpha(G) - 1$;

– if $V' \cap \Gamma'(v) \neq \emptyset$, then the independent set $S^* \setminus \{v\}$ (included in $G[V']$) can be augmented by exactly one element of $\Gamma'(v)$, so $|S_2^*| = \alpha(G)$.

We now study the two following cases with respect to $\Gamma'(v)$, namely $\Gamma'(v) = \emptyset$ and $\Gamma'(v) \neq \emptyset$.

$\Gamma'(v) = \emptyset$

Here, $|S_2^*| = \alpha(G) - 1$ and the following holds:

$$\sum_{\substack{V' \subseteq V \\ (S^* \setminus \{v\}) \subseteq V'}} \Pr[V'] \, |S_2^*| = (\alpha(G) - 1) \sum_{\substack{V' \subseteq V \\ (S^* \setminus \{v\}) \subseteq V'}} \Pr[V']$$

$$= p^{\alpha(G)-1}(1-p) \sum_{i=0}^{n-\alpha(G)} \binom{n-\alpha(G)}{i} p^i (1-p)^{n-\alpha(G)-i}$$

$$= p^{\alpha(G)-1}(1-p)$$

Consequently:

$$\sum_{\substack{V' \subseteq V \\ (S^* \setminus \{v\}) \subseteq V'}} \Pr[V'] \, |S_2^*| = (\alpha(G) - 1) \, p^{\alpha(G)-1}(1-p)$$

$\Gamma'(v) \neq \emptyset$

Here, we study the following two subcases: $V' \cap \Gamma'(v) = \emptyset$ and $V' \cap \Gamma'(v) \neq \emptyset$.

$V' \cap \Gamma'(v) = \emptyset$.

In this case, $|S_2^*| = \alpha(G) - 1$ and, moreover, no vertex of $\Gamma'(v)$ is contained in V'. We so get:

$$\sum_{\substack{V' \subseteq V \\ (S^* \setminus \{v\}) \subseteq V' \\ V' \cap \Gamma'(v) = \emptyset}} \Pr[V'] = p^{\alpha(G)-1}(1-p)(1-p)^{|\Gamma'(v)|}$$

$$= p^{\alpha(G)-1}(1-p)^{|\Gamma'(v)|+1} \qquad [2.36]$$

$V' \cap \Gamma'(v) \neq \emptyset$.
For the case covered: $|S_2^*| = \alpha(G)$ and V' contains at least one vertex of $\Gamma'(v)$; hence:

$$\sum_{\substack{V' \subseteq V \\ (S^* \setminus \{v\}) \subseteq V' \\ V' \cap \Gamma'(v) \neq \emptyset}} \Pr[V'] = p^{\alpha(G)-1}(1-p)\left[1-(1-p)^{|\Gamma'(v)|}\right] \quad [2.37]$$

Consequently, when $\Gamma'(v) \neq \emptyset$ we have combining expressions [2.36] and [2.37] together with expressions for $|S_2^*|$ of the two subcases:

$$\sum_{\substack{V' \subseteq V \\ (S^* \setminus \{v\}) \subseteq V'}} \Pr[V']\,|S_2^*| = p^{\alpha(G)-1}(1-p)\left[\alpha(G)-(1-p)^{|\Gamma'(v)|}\right] \quad [2.38]$$

This concludes the study of case $\Gamma'(v) \neq \emptyset$.

By summing [2.38] over all elements of S^*, we get:

$$\begin{aligned}
A_{\alpha(G)-1} &= \sum_{v \in S^*} \sum_{V' \subseteq V} \Pr[V']\,|S_2^*|\,1_{\{(S^* \setminus \{v\}) \subseteq V'\}} \\
&= \sum_{v \in S^*} \sum_{V' \subseteq V} \Pr[V']\,|S_2^*|\,1_{\{(S^* \setminus \{v\}) \subseteq V'\}} \\
&\quad \times \left(1_{\{\Gamma'(v)=\emptyset\}} + 1_{\{\Gamma'(v)\neq\emptyset\}}\right) \\
&= \ell_1\,(\alpha(G)-1)\,p^{\alpha(G)-1}(1-p) \\
&\quad + \sum_{\substack{v \in S^* \\ \Gamma'(v)\neq\emptyset}} p^{\alpha(G)-1}(1-p)\left(\alpha(G)-(1-p)^{|\Gamma'(v)|}\right) \\
&= \ell_1\,(\alpha(G)-1)\,p^{\alpha(G)-1}(1-p) \\
&\quad + (\alpha(G)-\ell_1)\,p^{\alpha(G)-1}(1-p)\alpha(G) \\
&\quad - p^{\alpha(G)-1}(1-p) \sum_{\substack{v \in S^* \\ \Gamma'(v)\neq\emptyset}} (1-p)^{|\Gamma'(v)|} \\
&= p^{\alpha(G)-1}(1-p) \\
&\quad \times \left(-\ell_1 + \alpha(G)^2 - \sum_{\substack{v \in S^* \\ \Gamma'(v)\neq\emptyset}} (1-p)^{|\Gamma'(v)|}\right)
\end{aligned}$$

$$= p^{\alpha(G)-1}(1-p) \times \left(-\ell_1 + \alpha(G)\left(\ell_1 + \ell_1'\right) - \sum_{\substack{v \in S^* \\ \Gamma'(v) \neq \emptyset}} (1-p)^{|\Gamma'(v)|} \right)$$

This concludes the proof of the proposition.

2.8.2. *Proof of Theorem 2.6*

The following expression holds for the expectation of s_i', $i = 1, \ldots, n - |S|$:

$$E\left(s_i'\right) = E\left(s_i' | v_i \text{ present}\right) p_i + E\left(s_i' | v_i \text{ absent}\right)\left(1 - p_i\right) \quad [2.39]$$

Moreover, for strategy M3 we have the following relation, setting $S_0' = \emptyset$, $\Gamma(S_0') = \emptyset$ and $s_0' = 0$:

$$s_i' = \begin{cases} s_{i-1}' & \text{if } \Gamma\left(v_i\right) \cap S_{i-1}' \neq \emptyset \\ s_{i-1}' + 1 & \text{otherwise} \end{cases} \quad [2.40]$$

So, using [2.40], we get: $E(s_i' | v_i \text{ present}) = E(s_{i-1}') + \Pr[\Gamma(v_i) \cap S_{i-1}' = \emptyset]$; consequently, using also [2.39]:

$$E\left(s_i'\right) = p_i E\left(s_{i-1}'\right) + p_i \Pr\left[v_i \notin \Gamma\left(S_{i-1}'\right)\right] + \left(1 - p_i\right) E\left(s_{i-1}'\right)$$

Since $E(s_0') = 0$:

$$E\left(s_i'\right) = \sum_{j=1} p_j \Pr\left[v_j \notin \Gamma\left(S_{j-1}'\right)\right]$$

$$E\left(s_{n-|S|}'\right) = \sum_{j=1}^{n-|S|} p_i \Pr\left[v_i \notin \Gamma\left(S_{i-1}'\right)\right]$$

Now, let $V'' = f(V') = V' \setminus \{S(V') \cup \Gamma(S(V'))\}$ and let $s(G[V''])$ be the cardinality of the solution provided by M3 when applied in graph $G[V'']$, $V'' \subseteq V$.

This set represents the subset of vertices of V' which are contained neither in S, nor in the neighbor-set of $S(V')$. Consequently, M3 will be applied on $G[V'']$; so $|S_3'| = |S(V')| + s(G[V''])$ and, consequently:

$$E(G,S,\mathtt{M3}) = \sum_{V'\subseteq V} \Pr\left[V'\right]\left|S\left(V'\right)\right| + \sum_{V'\subseteq V} \Pr\left[V'\right] s\left(G\left[V''\right]\right)$$

$$= E(G,S,\mathtt{M1}) + \sum_{V'\subseteq V} \Pr\left[V'\right] s\left(G\left[V''\right]\right)$$

$$= E(G,S,\mathtt{M1}) + \sum_{V'\subseteq V} \left(\sum_{V''\subseteq V} 1_{\{f(V')=V''\}}\right) \Pr\left[V'\right] s\left(G\left[V''\right]\right)$$

$$= E(G,S,\mathtt{M1}) + \sum_{V''\subseteq V} \sum_{\substack{V'\subseteq V\\ f(V')=V''}} \Pr\left[V'\right] s\left(G\left[V''\right]\right)$$

Since $\Pr[V''] = \sum_{V'\subseteq V, f(V')=V''} \Pr[V']$, we get:

$$E(G,S,\mathtt{M3}) = E(G,S,\mathtt{M1}) + \sum_{V''\subseteq V} \Pr\left[V''\right] s\left(G\left[V''\right]\right)$$

$$= E(G,S,\mathtt{M1}) + E\left({s'}_{n-|S|}\right)$$

Let us now introduce the random variable C_i representing the solution of PROBABILISTIC MAX INDEPENDENT SET3 whenever we consider only the present vertices of V_i, i.e., when we apply the greedy algorithm implied by strategy M3 in the graph induced by the present vertices of V_i:

$$C_i = \begin{cases} C_{i-1} & v_i \text{ is absent} \\ C_{i-1} & v_i \text{ is present and } \Gamma\left(v_i\right) \cap C_{i-1} \neq \emptyset \\ C_{i-1} \cup \{v_i\} & v_i \text{ is present and } \Gamma\left(v_i\right) \cap C_{i-1} = \emptyset \end{cases}$$

We then have $E(s'_{n-|S|}) = \sum_{i=1}^{n-|S|} p_i \Pr[\Gamma(v_i) \cap C_{i-1} = \emptyset]$.

In order to prove the result of the theorem, we prove that the quantity $\Pr[\Gamma(v_i) \cap C_{i-1} = \emptyset]$ is not computable in polynomial time. Indeed:

$$\Pr\left[\Gamma\left(v_i\right) \cap C_{i-1} = \emptyset\right] = \left(1 - p_{i-1}\right) \Pr\left[\Gamma\left(v_i\right) \cap C_{i-2} = \emptyset\right]$$

$$
\begin{aligned}
+ \quad & (p_{i-1} \Pr[\Gamma(v_i) \cap C_{i-2} = \emptyset] \Pr[\Gamma(v_{i-1}) \cap C_{i-2} \neq \emptyset]) \\
& + (p_{i-1} \Pr[\Gamma(v_i) \cap (C_{i-2} \cup \{v_{i-1}\}) = \emptyset] \Pr[\Gamma(v_{i-1}) \cap C_{i-2} = \emptyset]) \\
= \quad & (1 - p_{i-1}) \Pr[\Gamma(v_i) \cap C_{i-2} = \emptyset] \\
& + (p_{i-1} \Pr[\Gamma(v_i) \cap C_{i-2} = \emptyset] (1 - \Pr[\Gamma(v_{i-1}) \cap C_{i-2} = \emptyset])) \\
& + (p_{i-1} \Pr[\Gamma(v_i) \cap (C_{i-2} \cup \{v_{i-1}\}) = \emptyset] \Pr[\Gamma(v_{i-1}) \cap C_{i-2} = \emptyset]) \\
= \quad & \Pr[\Gamma(v_i) \cap C_{i-2} = \emptyset] \\
& - (p_{i-1} \Pr[\Gamma(v_i) \cap C_{i-2} = \emptyset] \Pr[\Gamma(v_{i-1}) \cap C_{i-2} = \emptyset]) \\
& + (p_{i-1} \Pr[\Gamma(v_i) \cap (C_{i-2} \cup \{v_{i-1}\}) = \emptyset] \Pr[\Gamma(v_{i-1}) \cap C_{i-2} = \emptyset])
\end{aligned}
$$

On the other hand:

$$
\begin{aligned}
& \Pr[\Gamma(v_i) \cap (C_{i-2} \cup \{v_{i-1}\}) = \emptyset] = \\
& \qquad \Pr[(\Gamma(v_i) \cap C_{i-2}) \cup (\Gamma(v_i) \cap \{v_{i-1}\}) = \emptyset] \\
& = \quad \Pr[(\Gamma(v_i) \cap C_{i-2} = \emptyset) \cap (\Gamma(v_i) \cap \{v_{i-1}\} = \emptyset)] \\
& = \quad \Pr[\Gamma(v_i) \cap C_{i-2} = \emptyset] \, 1_{\{\Gamma(v_i) \cap \{v_{i-1}\} = \emptyset\}}
\end{aligned}
$$

Consequently:

$$
\begin{aligned}
& \Pr[\Gamma(v_i) \cap C_{i-1} = \emptyset] = \Pr[\Gamma(v_i) \cap C_{i-2} = \emptyset] \\
& \quad - (p_{i-1} \Pr[\Gamma(v_i) \cap C_{i-2} = \emptyset] \Pr[\Gamma(v_{i-1}) \cap C_{i-2} = \emptyset]) \\
& \quad + (p_{i-1} \Pr[\Gamma(v_i) \cap C_{i-2} = \emptyset] \Pr[\Gamma(v_{i-1}) \cap C_{i-2} = \emptyset] \\
& \quad \times 1_{\{\Gamma(v_i) \cap \{v_{i-1}\} = \emptyset\}})
\end{aligned}
$$

Let $t_{i-1}(v_i)$ be the computational time of $\Pr[\Gamma(v_i) \cap C_{i-1} = \emptyset]$. By the above equalities we easily deduce that:

$$
t_{i-1}(v_i) = t_{i-2}(v_i) + t_{i-2}(v_{i-1}) \qquad [2.41]
$$

In order to compute recurrence relation given in [2.41], at each step we need to know two terms of the precedent step; so for the computation of $\Pr[\Gamma(v_i) \cap C_{i-1} = \emptyset]$, we need 2^{i-1} computations and expression for $T(E(G, S, \mathtt{M3}))$ immediately follows. The proof of the theorem is now complete.

2.8.3. *Proof of Proposition 2.3*

Set $B_i = \Pr[X_i^{\mathcal{U}4,S} = 1]$. Then

$$B_i = \sum_{V' \subseteq V} \Pr[V'] \, 1_{\{v_i \in S'_4\}}$$

Let v_i be any vertex of $V \setminus S$ and let V' be any subset of V containing v_i. Obviously:

$$\begin{aligned} v_i \in S'_4 \quad &\Longleftrightarrow \quad v_i \text{ isolated in } G\left[\tilde{V}'\right] \\ &\Longleftrightarrow \quad \left(v_i \in G\left[\tilde{V}'\right]\right) \wedge \left(v_i \text{ has lost all its neighbors in } \tilde{V}'\right) \end{aligned}$$

Since $v_i \notin S$, $v_i \in \tilde{V}'$ only if v_i does not belong to the neighborhood of any vertex in $S(V')$, i.e.:

$$v_i \in G\left[\tilde{V}'\right] \Longleftrightarrow \Gamma_S(v_i) \cap V' = \emptyset \qquad [2.42]$$

On the other hand, all the neighbors of v_i in $G[\tilde{V}']$ have been removed if and only if $\Gamma_{V \setminus S}(v_i) \cap \tilde{V}' = \emptyset$. This last condition is satisfied only if every vertex of $\Gamma_{V \setminus S}(v_i)$ is either absent, or (being present) has been removed from $\tilde{V}'$ because it belonged to the neighborhood of a vertex in $S(V')$. In all:

$$\begin{aligned} \Gamma_{V \setminus S}(v_i) \cap \tilde{V}' = \emptyset \quad \Longleftrightarrow \quad & \forall v_j \in \Gamma_{V \setminus S}(v_i) \left((v_j \text{ is absent }) \vee \right. \\ & ((v_j \text{ is present}) \wedge \\ & \left. (\exists v_k \in \Gamma(v_j) \cap S : v_k \text{ is present})) \right) \end{aligned} \qquad [2.43]$$

From [2.42] and [2.43] and the discussion above, we get (assuming that $v_i \in V \setminus S$):

$$\begin{aligned} B_i &= \sum_{V' \subseteq V} \Pr[V'] \, 1_{\{v_i \in S'_4\}} = \sum_{V' \subseteq V} \Pr[V'] \, 1_{\left\{ \substack{v_i \in G[\tilde{V}'] \\ \Gamma_{V \setminus S}(v_i) \cap \tilde{V}' = \emptyset} \right\}} \\ &= \sum_{V' \subseteq V} \Pr[V'] \, 1_{\{v_i \in G[\tilde{V}']\}} 1_{\{\Gamma_{V \setminus S}(v_i) \cap \tilde{V}' = \emptyset\}} \\ &= \sum_{V' \subseteq V} \Pr[V'] \, 1_{\left\{ \substack{\{v_i\} \cap V' = \{v_i\} \\ \Gamma_S(v_i) \cap V' = \emptyset} \right\}} 1_{\{\Gamma_{V \setminus S}(v_i) \cap \tilde{V}' = \emptyset\}} \end{aligned}$$

$$= \sum_{V' \subseteq V} \Pr[V'] \, 1_{\{\{v_i\} \cap V' = \{v_i\}\}} 1_{\{\Gamma_S(v_i) \cap V' = \emptyset\}}$$

$$\times \prod_{v_j \in \Gamma_{V \setminus S}(v_i)} \left(1_{\{V' \cap \{v_j\} = \emptyset\}} + 1_{\{V' \cap \{v_j\} = \{v_j\}\}} 1_{\{V' \cap \Gamma_S(v_j) \neq \emptyset\}}\right)$$

$$= p_i \prod_{v_k \in \Gamma_S(v_i)} (1 - p_k)$$

$$\times \prod_{v_j \in \Gamma_{V \setminus S}(v_i)} \left((1 - p_j) + p_j \left(1 - \prod_{v_l \in \Gamma_S(v_j)} (1 - p_l)\right)\right) \quad [2.44]$$

Replacing the second term of [2.2] by [2.44], we easily obtain [2.18].

Moreover, one can see that computation of $E(G, S, \mathtt{M4})$ in [2.18] takes at most n^2 multiplications. Also, setting $p_i = p$, $\forall v_i \in V$, we immediately obtain [2.19].

2.8.4. *Proof of Theorem 2.13*

Set $|\Gamma_S(v_i)| = k_i$, $|\Gamma_{V \setminus S}(v_i)| = |\Gamma(v_i)| - k_i$. Then, [2.19] becomes:

$$E(G, S, \mathtt{M4}) = p|S| + \sum_{v_i \in (V \setminus S)} p(1-p)^{k_i} \times \prod_{v_k \in \Gamma_{V \setminus S}(v_i)} \left(1 - p(1-p)^{k_k}\right) \quad [2.45]$$

Also, note that $\forall k$, $k_k \leqslant \Delta(G) - 1$ (v_i and v_k are in $V \setminus S$ and $v_i v_k \in E$) and $E(G, S, \mathtt{M4})$ is increasing in k_k. Then, using [2.45]:

$$E\left(G, \hat{S}, \mathtt{M4}\right) \leqslant p\left|\hat{S}\right| \quad [2.46]$$

$$+ \left(\sum_{v_i \in (V \setminus \hat{S})} p(1-p)^{k_i} \prod_{v_k \in \Gamma_{V \setminus \hat{S}}(v_i)} \left(1 - p(1-p)^{\Delta(G)-1}\right)\right)$$

$$\leqslant p\left|\hat{S}\right| + \left(\sum_{v_i \in (V \setminus \hat{S})} p(1-p)^{k_i} \left(1 - p(1-p)^{\Delta(G)-1}\right)^{|\Gamma(v_i)| - k_i}\right)$$

$$\leqslant p\left|\hat{S}\right| + \sum_{v_i \in \left(V \setminus \hat{S}\right)} p\left(1 - p(1-p)^{\Delta(G)-1}\right)^{|\Gamma(v_i)|} \quad [2.47]$$

$$\leqslant p\left|\hat{S}\right| + \left(n - \left|\hat{S}\right|\right) p\left(1 - p(1-p)^{\Delta(G)-1}\right) \leqslant pn \quad [2.48]$$

where in [2.47] is used the fact that $1 - p \leqslant 1 - p(1-p)^{\Delta(G)-1}$ and in [2.48] the fact that $|\Gamma(v_i)| \geqslant 1$.

On the other hand, if we consider S^* as an *a priori* solution, we get:

$$E\left(G, S^*, \texttt{M4}\right) = p\alpha(G) + \left(\sum_{v_i \in (V \setminus S^*)} p(1-p)^{k_i} \prod_{v_k \in \Gamma_{V \setminus S^*}(v_i)} \left(1 - p(1-p)^{k_k}\right)\right)$$

$$\geqslant p\alpha(G) + \left(\sum_{v_i \in (V \setminus S^*)} p(1-p)^{k_i} \prod_{v_k \in \Gamma_{V \setminus S^*}(v_i)} (1-p)\right) \quad [2.49]$$

$$= p\alpha(G) + \sum_{v_i \in (V \setminus S^*)} p(1-p)^{k_i}(1-p)^{|\Gamma(v_i)| - k_i}$$

$$= p\alpha(G) + \sum_{v_i \in (V \setminus S^*)} p(1-p)^{|\Gamma(v_i)|}$$

$$\geqslant p\alpha(G) + (n - \alpha(G))\, p(1-p)^{\Delta(G)}$$

$$= p\alpha(G)\left(1 - (1-p)^{\Delta(G)}\right) + pn(1-p)^{\Delta(G)} \quad [2.50]$$

where in [2.49] the fact is used that $1 - p(1-p)^{k_k} \geqslant 1 - p$.

Combining expressions [2.48] and [2.50], we get:

$$\frac{E\left(G, S^*, \texttt{M4}\right)}{E\left(G, \hat{S}, \texttt{M4}\right)} \geqslant \frac{\alpha(G)}{n}\left(1 - (1-p)^{\Delta(G)}\right) + (1-p)^{\Delta(G)} \geqslant \frac{\alpha(G)}{n} \geqslant \frac{1}{\Delta(G)+1}$$

that concludes the proof of the theorem.

Chapter 3

The Probabilistic Minimum Vertex Cover

This chapter is complementary to Chapter 2 since we study the probabilistic vertex covering problem that is, as we have seen in Chapter 2 (see also Appendix A), complementary to the probabilistic independent set problem. Despite this complementarity (and simultaneously because of it), the results that we present have their own interest.

Given a graph $G(V, E)$ together with a probability-vector associating a presence probability $\Pr[v_i] = p_i$ with any vertex $v_i \in V$ and a modification strategy $\mathtt{M}$, the objective for MIN PROBABILISTIC VERTEX COVER is to determine:

$$\hat{C} = \underset{C \in \mathsf{C}(G)}{\operatorname{argmin}}\{E(G, C, \mathtt{M})\}$$

where $\mathsf{C}(G)$ denotes the set of vertex covers of G, and the functional $E(G, C, \mathtt{M})$ is expressed as:

$$E(G, C, \mathtt{M}) = \sum_{V' \subseteq V} \Pr[V'] \, |C(V', \mathtt{M})|$$

with $\Pr[V'] = \prod_{i \in V'} p_i \prod_{i \in V \setminus V'} (1 - p_i)$.

In this chapter we devise three modification strategies denoted by $\mathtt{M1}$, $\mathtt{M2}$ and $\mathtt{M3}$, and study their corresponding functionals. For $\mathtt{M1}$ and $\mathtt{M2}$, we produce explicit expressions for their functionals, expressions allowing us to completely characterize the solutions minimizing these functionals. On the contrary, the expression for the functional associated with $\mathtt{M3}$ is not quite explicit in order to allow achievement of similar results. Therefore, we give bounds for this functional and study their quality. We recall once more that the three strategies in fact introduce three distinct probabilistic PROBABILISTIC MIN VERTEX COVER-variants, denoted in the sequel by PROBABILISTIC

MIN VERTEX COVER1, PROBABILISTIC MIN VERTEX COVER2 and PROBABILISTIC MIN VERTEX COVER3, respectively.

In what follows, given a graph $G(V, E)$ of order n, we denote by C a vertex cover of G and by $\hat{C}_i$, $i = 1, 2, 3$ the optimal PROBABILISTIC MIN VERTEX COVER-solutions associated with M1, M2 and M3, respectively. Given a vertex cover C of G and a subset $V' \subseteq V$, we will set $C(V') = C \cap V'$ and $\bar{C}(V') = (V \setminus C) \cap V'$. Finally, when dealing with MIN WEIGHTED VERTEX COVER, we will denote by w_i the weight of $v_i \in V$.

3.1. The strategies M1, M2 and M3 and a general preliminary result

3.1.1. *Specification of* M1, M2 *and* M3

3.1.1.1. *Strategy M1*

Given a vertex cover C and a vertex-subset $V' \subseteq V$, strategy M1 consists of simply moving vertices of $C \setminus V'$ (the absent vertices of C) out of C and of retaining $C'_1 = C(V') = C \cap V'$ as vertex cover of $G[V']$. In other words, M1 is the strategy (dealing with PROBABILISTIC VERTEX COVER) denoted by MS in Chapter 1.

It is easy to see that C'_1 constitutes a vertex cover for $G[V']$; in the opposite case, there would be at least an edge (v_i, v_j) of $G[V']$ for which neither v_i nor v_j would belong to C'_1. But, since $(v_i, v_j) \in E$ and C is a vertex cover for G, at least one of v_i, v_j, say v_i, belongs to C and, consequently, to C'_1. Therefore v_i is part of the vertex cover for $G[V']$, thus contradicting the statement that edge (v_i, v_j) is not covered in $G[V']$.

EXAMPLE 3.1.– Consider the graph of Figure 2.1, the *a priori* vertex cover $C = \{2, 4, 5, 6, 9\}$ and assume that vertices 3, 5 and 8 are absent. Then strategy M1 will produce a vertex cover $C'_1 = \{2, 4, 6, 9\}$ (the black vertices of Figure 3.1).

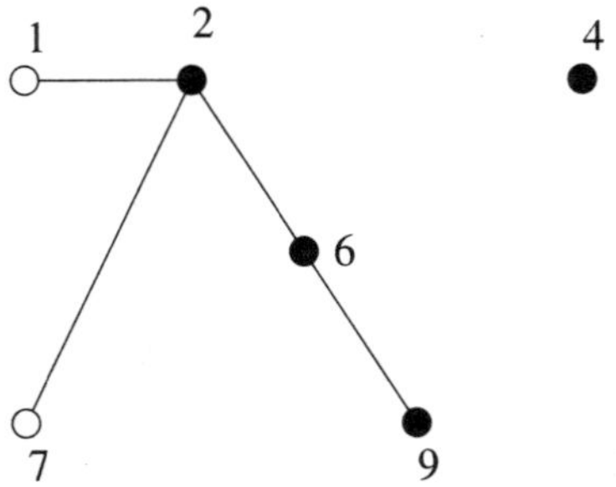

Figure 3.1. *Application of strategy M1 on G', with a priori solution C ("black" vertices of Figure 2.1); vertex cover produced: $\{2, 4, 6, 9\}$*

3.1.1.2. *Strategy M2*

The second strategy studied in this paper is a slight improvement of `M1` since one removes isolated vertices from C_1'. It works as follows:

– set: $C_1' \leftarrow C \cap V'$, $R = \{v_i \in C_1' : \Gamma(v_i) \cap V' = \emptyset\}$;

– output $C_2' = C_1' \setminus R$.

EXAMPLE 3.2.– Consider again the data of Example 3.1. The first item of strategy `M2` will produce $C_1' = \{2, 4, 6, 9\}$ and $R = \{4\}$. Then, the second item of `M2` will return $C_2' = \{2, 6, 9\}$ (Figure 3.2).

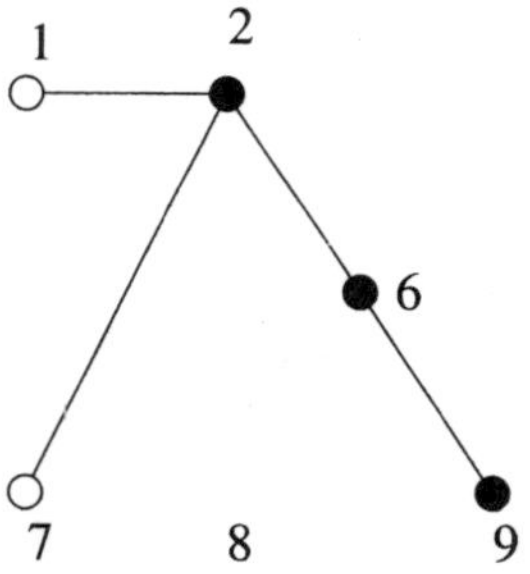

Figure 3.2. *Application of strategy* `M2` *on* G', *with a priori solution* C *("black" vertices of Figure 2.1); vertex cover produced:* $\{2, 6, 9\}$

3.1.1.3. *Strategy M3*

Strategy `M3` is a further improvement of `M2` since it removes from C_2' vertices all the neighbors of which belong to C_2'. In all, strategy `M3` works as follows:

– set $C_2' = C(G[V'], \texttt{M2})$;

– range the vertices in C_2' in increasing order with respect to their degrees; let C_3' be the cover so obtained;

– for $i = 1$ to $|C_3'|$: if, for every $v_j \in \Gamma(v_i) \cap V'$, $v_j \in C_3'$, then set $C_3' = C_3' \setminus \{v_i\}$;

– output C_3'.

It is easy to see that the vertex cover C_3' computed by strategy `M3` is minimal (for the inclusion) for $G[V']$.

EXAMPLE 3.3.– Consider once more the data of Example 3.1. The first item of strategy `M3` will produce $C_2' = \{2, 6, 9\}$, while the third item will drop 9 out of C_2'. Hence, $C_3' = \{2, 6\}$ (Figure 3.3).

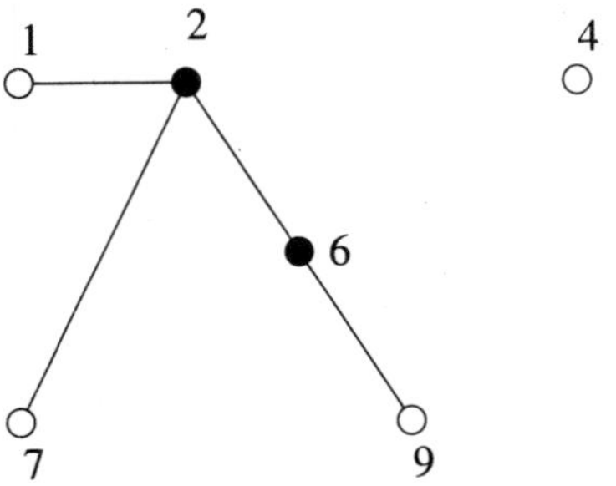

Figure 3.3. *Application of strategy M3 on G', with a priori solution C ("black" vertices of Figure 2.1); vertex cover produced:* $\{2, 6\}$

3.1.2. *A first expression for the functionals*

We establish in this section a general result about the functional. This result is analogous to the one in Theorem 2.1 of Chapter 2.

Consider any strategy where starting from C'_1 reduces it by removing some of its vertices (if possible) in order to obtain smaller feasible vertex covers. Clearly, M1, M2 and M3 are such strategies. Then, the following proposition provides a first expression for functionals $E(G, C, \mathtt{M1})$, $E(G, C, \mathtt{M2})$ and $E(G, C, \mathtt{M3})$.

PROPOSITION 3.1.– *Consider an a priori vertex cover C of G and strategies M1, M2 and M3. With each vertex $v_i \in V$ associate a probability p_i and a random variable $X_i^{\mathtt{Mk},C}$, $k = 1, 2, 3$, defined, for any $V' \subseteq V$, by:*

$$X_i^{\mathtt{Mk},C} = \begin{cases} 1 & \text{if } v_i \in C\left(G\left[V'\right], \mathtt{Mk}\right) \\ 0 & \text{otherwise} \end{cases}$$

Then, $E(G, C, \mathtt{Mk}) = \sum_{v_i \in C} \Pr[X_i^{\mathtt{Mk},C} = 1]$.

Proof. The linearity of the functional implies:

$$E(G, C, \mathtt{Mk}) = E\left[\sum_{v_i \in C} X_i^{\mathtt{Mk},C}\right] = \sum_{v_i \in C} \Pr\left[X_i^{\mathtt{Mk},C} = 1\right]$$

that concludes the proof. ∎

3.2. PROBABILISTIC MIN VERTEX COVER1

The following result, characterizing the complexity of PROBABILISTIC MIN VERTEX COVER1, can be deduced by a straightforward application of Proposition 3.1.

THEOREM 3.1.– *$E(G, C, \mathtt{M1}) = \sum_{v_i \in C} p_i$ and $\hat{C}_1 = \operatorname{argmin}_{C \in \mathsf{C}(G)}\{\sum_{v_i \in C} p_i\}$. In other words, $\hat{C}_1$ is a minimum-weight vertex cover of G where the vertices of V are weighted by their corresponding presence-probabilities. In the case of identical vertex-probabilities, $E(G, C, \mathtt{M1}) = p|C|$ and $\hat{C}_1$ is a minimum vertex cover of G. Consequently,* PROBABILISTIC MIN VERTEX COVER*1 is **NP**-hard.*

Proof. By strategy M1, if $v_i \in C$, then $v_i \in C_1'$, for all subgraphs $G[V']$ such that $v_i \in V'$. Consequently, $\Pr[X_i^{\mathtt{M1};C} = 1] = p_i$, $\forall v_i \in C$ and the result follows from Proposition 3.1.

Let us now consider G with its vertices weighted by their corresponding probabilities and denote the so obtained instance of MIN WEIGHTED VERTEX COVER by G_w. The total weight of every vertex cover of G_w is $\sum_{v_i \in C} p_i$ and the optimal weighted vertex cover of G_w is the one for which the sum of the weights of its vertices is the smallest over all vertex covers of G_w. Such a vertex cover also minimizes $E(G, C, \mathtt{M1})$ and, consequently, constitutes an optimal solution for PROBABILISTIC MIN VERTEX COVER1.

If $p_i = p$, $1 \leqslant i \leqslant n$, then $\sum_{v_i \in C} p_i = \sum_{v_i \in C} p = p|C|$ and, consequently, the vertex cover minimizing this last expression is the one minimizing $|C|$, i.e., a minimum vertex cover of G.

Expression $\sum_{v_i \in C} p_i$ (which represents the objective value of PROBABILISTIC MIN VERTEX COVER1) can be computed in $O(n)$ and, therefore, PROBABILISTIC MIN VERTEX COVER1 belongs to **NPO**. Furthermore, by the correspondence between this problem and MIN WEIGHTED VERTEX COVER, the former is hard for **NPO** and this completes the proof of Theorem 3.1. ∎

Let us revisit the case $p_i = p$, $1 \leqslant i \leqslant n$. Since $\sum_{v_i \in C} X_i^{\mathtt{M1},C}$ can be seen either as a binomial random variable, or as the sum of $|C|$ Bernoulli random variables, then:

$$\begin{aligned} E(G, C, \mathtt{M1}) &= E\left[\sum_{v_i \in C} X_i^{\mathtt{M1},C}\right] = p|C| \\ \operatorname{Var}(G, C, \mathtt{M1}) &= \operatorname{Var}\left(\sum_{v_i \in C} X_i^{\mathtt{M1},C}\right) = p(1-p)|C| \end{aligned}$$

So, as in Chapter 2, dealing with M1 and identical vertex-probabilities, the random variable representing the size $|C|$ follows a binomial law with parameters $|C|$ and p.

3.3. PROBABILISTIC MIN VERTEX COVER2

We recall that strategy M2 consists in removing the isolated vertices from C_1'. Then, the following theorem holds.

THEOREM 3.2.– *The functional* $E(G, C, \texttt{M2})$ *of* PROBABILISTIC MIN VERTEX COVER2 *can be expressed as:*

$$E(G, C, \texttt{M2}) = \sum_{v_i \in C} p_i \left(1 - \prod_{v_j \in \Gamma(v_i)} (1 - p_j)\right)$$

$\hat{C}_2$ *is a minimum-weight vertex cover of* G_w *where, for every* $v_i \in V$*:*

$$w_i = p_i \left(1 - \prod_{v_j \in \Gamma(v_i)} (1 - p_j)\right)$$

Consequently, PROBABILISTIC MIN VERTEX COVER2 *is* ***NP****-hard.*

Proof. Note that $\Pr[X_i^{\texttt{M2},C} = 1]$ is just the probability that v_i is present and at least one of its neighbors is also present; consequently:

$$\Pr\left[X_i^{\texttt{M2},C} = 1\right] = p_i \left(1 - \prod_{v_j \in \Gamma(v_i)} (1 - p_j)\right) \quad [3.1]$$

Thus, starting from Proposition 3.1 we get:

$$E(G, C, \texttt{M2}) = \sum_{v_i \in C} \Pr\left[X_i^{\texttt{M2},C} = 1\right] = \sum_{v_i \in C} p_i \left(1 - \prod_{v_j \in \Gamma(v_i)} (1 - p_j)\right)$$

It is easy to see that $E(G, C, \texttt{M2})$ can be computed in $O(n^2)$; consequently, PROBABILISTIC MIN VERTEX COVER2 is in **NPO**. With a reasoning completely similar to the one of Theorem 3.1, one can immediately deduce that:

$$\hat{C}_2 = \underset{C \in C(G)}{\operatorname{argmin}} \left\{ \sum_{v_i \in C} p_i \left(1 - \prod_{v_j \in \Gamma(v_i)} (1 - p_j)\right) \right\}$$

i.e., a minimum-weight vertex cover of G_w where, for $v_i \in V$:

$$w_i = p_i \left(1 - \prod_{v_j \in \Gamma(v_i)} (1 - p_j)\right)$$

The only characteristic of numbers w_i is that they are all smaller than 1.

Consider an instance $(G, \vec{w})$ of MIN WEIGHTED VERTEX COVER, where vertex weights are all greater than 1. Divide them by, say M, a number greater than the maximum-value component of $\vec{w}$. We thus obtain an instance $(G, \vec{w}')$ where, obviously, all components of $\vec{w}'$ are smaller than 1. It is easy to see that every weighted vertex cover of $(G, \vec{w}')$ of value W' can be directly transformed into a weighted vertex cover of $(G, \vec{w})$ of value MW'. Conversely, every weighted vertex cover of $(G, \vec{w})$ of value W can be directly transformed into a weighted vertex cover of $(G, \vec{w}')$ of value W/M. Therefore, PROBABILISTIC MIN VERTEX COVER2 is **NP**-hard, and the proof of the theorem is complete. ∎

The characterizations for the *a priori* solutions for PROBABILISTIC MIN VERTEX COVER1 and PROBABILISTIC MIN VERTEX COVER2 given by Theorems 3.1 and 3.2, respectively, allow us to derive immediate approximation results for them. Indeed, MIN WEIGHTED VERTEX COVER is approximable within ratio bounded above by 2 ([AUS 99, BAR 85, PAS 97]). Since both problems are weighted vertex cover versions, they can also be approximately solved within ratio 2.

3.4. PROBABILISTIC MIN VERTEX COVER3

3.4.1. *Building* $E(G, C, \mathtt{M3})$

Recall that M3 consists of removing from C_1' both the isolated vertices and the vertices all the neighbors of which belong to C_1'. Consequently, the solution $C_3(V')$ computed by M3 is minimal.

PROPOSITION 3.2.– *The functional of* PROBABILISTIC MIN VERTEX COVER3 *can be expressed as:*

$$\begin{aligned} E(G, C, \mathtt{M3}) &= \sum_{v_i \in C} p_i \left(1 - \prod_{v_j \in \Gamma(v_i)} (1 - p_j)\right) \\ &\quad - \sum_{v_i \in C} \sum_{\substack{V' \subseteq V \\ v_i \in V'}} \Pr\left[V'\right] 1_{\{v_i \notin C_3\}} 1_{\{V' \cap C[\Gamma(v_i)] \neq \emptyset\}} \\ &\quad \times 1_{\{V' \cap \bar{C}[\Gamma(v_i)] = \emptyset\}} \end{aligned} \qquad [3.2]$$

Proof. Since $X_i^{\mathtt{M3},C} = 1$ implies $X_i^{\mathtt{M2},C} = 1$, we have:

$$\Pr\left[X_i^{\mathtt{M3},C} = 1\right] = \Pr\left[X_i^{\mathtt{M2},C} = 1\right] - \Pr\left[X_i^{\mathtt{M2},C} = 1 \wedge X_i^{\mathtt{M3},C} = 0\right] \qquad [3.3]$$

We now calculate $\Pr[X_i^{\mathtt{M2},C} = 1 \wedge X_i^{\mathtt{M3},C} = 0]$. $X_i^{\mathtt{M2},C} = 1$ implies that $\Gamma(v_i) = C[\Gamma(v_i)] \cup \bar{C}[\Gamma(v_i)]$ must intersect V'. Then, $X_i^{\mathtt{M2},C} = 1 \wedge X_i^{\mathtt{M3},C} = 0$ implies $V' \cap \bar{C}[\Gamma(v_i)] = \emptyset$ and $V' \cap C[\Gamma(v_i)] \neq \emptyset$. So:

$$\Pr\left[X_i^{\mathtt{M2},C} = 1 \wedge X_i^{\mathtt{M3},C} = 0\right] =$$
$$\sum_{\substack{V' \subseteq V \\ v_i \in V'}} \Pr[V'] \, 1_{\{v_i \notin C_3'\}} 1_{\{V' \cap C[\Gamma(v_i)] \neq \emptyset\}} 1_{\{V' \cap \bar{C}[\Gamma(v_i)] = \emptyset\}} \qquad [3.4]$$

From Proposition 3.1 and [3.1], [3.3] and [3.4], the proposition follows. ∎

3.4.2. *Bounds for* $E(G, C, \mathtt{M3})$

Let us first note that from Theorems 3.1, 3.2 and Proposition 3.2, we have:

$$E(G, C, \mathtt{M3}) \leqslant E(G, C, \mathtt{M2}) \leqslant E(G, C, \mathtt{M1})$$

Since the result of Proposition 3.2 does not allow the achievement of a precise characterization of $\hat{C}_3$, we give in the following theorem bounds (computable in polynomial time) for $E(G, C, \mathtt{M3})$. The proof of this theorem can be found in section 3.6.1.

THEOREM 3.3.– *The following inequalities hold for* $E(G, C, \mathtt{M3})$*:*

$$E(G, C, \mathtt{M3}) \geqslant \sum_{v_i \in C} p_i \left(1 - \prod_{v_j \in \bar{C}[\Gamma(v_i)]} (1 - p_j)\right)$$
$$E(G, C, \mathtt{M3}) \leqslant \sum_{v_i \in C} p_i \left(1 - \prod_{v_j \in \Gamma(v_i)} (1 - p_j)\right)$$

If $p_i = p$, $\forall v_i \in V$ *and* $\delta_C = \min_{v_i \in C}\{|\Gamma(v_i)|\}$, *then:*

$$p\left(|C| - \max\left\{|C|(1-p)^{\Delta(G)}, (1-p)^{\delta_C}\right\}\right) \geqslant E(G, C, \mathtt{M3}) \geqslant p^2|C|$$

The upper bound for $E(G, C, \mathtt{M3})$ *is attained for the bipartite graphs when considering one of the color classes as an a priori solution.*

3.5. Some methodological questions

As in Chapter 2, we pose here also some methodological questions dealing with the use of the expectation as measure for PROBABILISTIC MIN VERTEX COVER.

What does "minimizing the expectation" mean? In order to minimize the functional, we have rather to choose vertices of weak probability, i.e., vertices very probably absent, and, in this case, we construct solutions including elements unlikely to be present. For PROBABILISTIC MIN VERTEX COVER, in order to minimize the expectation of a cover C, i.e., $\min_C \sum_{v_i \in V} 1_{\{v_i \in C\}} p_i$, an eventual algorithm should try to introduce in the solution vertices of weak p_i. On the other hand, if a vertex v_i has small presence probability, then it is very likely that all of its incident edges will be absent even if their second endpoint has large probability to be present (we always admit the hypothesis that the modification strategy includes a step of removing the absent vertices and the edges incident to these vertices). Consequently, in order to cover these "improbable" edges, using small probability vertices (thus keeping the functional's value low) is not senseless. However, a solution constituted in great majority from small probability vertices is no more meaningful.

Of course, it is possible to try to overcome these ambiguities by defining other types of functionals. For example, one can try to minimize the expected size of a solution, all the elements of which have probabilities greater than a given threshold probability p_0. For example, consider objective:

$$\min_C \sum_{v_i \in V} 1_{\{v_i \in C : p_i \geqslant p_0\}} \qquad [3.5]$$

With this kind of functional, we answer, eventually, the problem of the functional's meaning, but other problems arise. For instance, have all of the edges at least one endpoint of strong probability? If this is not the case, then some edges may remain uncovered (we are thus faced with unfeasible solutions). Covering them by adding (perhaps greedily) low probability vertices, we build an artificial functional since these last vertices will not be taken into account in [3.5]. On the other hand, one can observe that there exists another problem with this expression, because p_is' values do not actively intervene in the functional's value. One can then propose a modification, for example $\min_C \sum_{v_i \in V} 1_{\{v_i \in C : p_i \geq p_0\}} p_i$, but even in this case, the drawbacks mentioned above always persist.

Let us show that alternative ways to define functional for PROBABILISTIC VERTEX COVER do not lead to identical results. In what follows, we do not restrict ourselves to the case of *a priori* optimization.

Consider the graph $G(V, E)$ of Figure 3.4. With any vertex $v_i \in V$, we associate probability p_i, $i = 1, \ldots, 14$. Set: $p_1 = 0.1$, $p_2 = 0.3$, $p_3 = 0.3$, $p_4 = 0.6$, $p_5 = 0.7$,

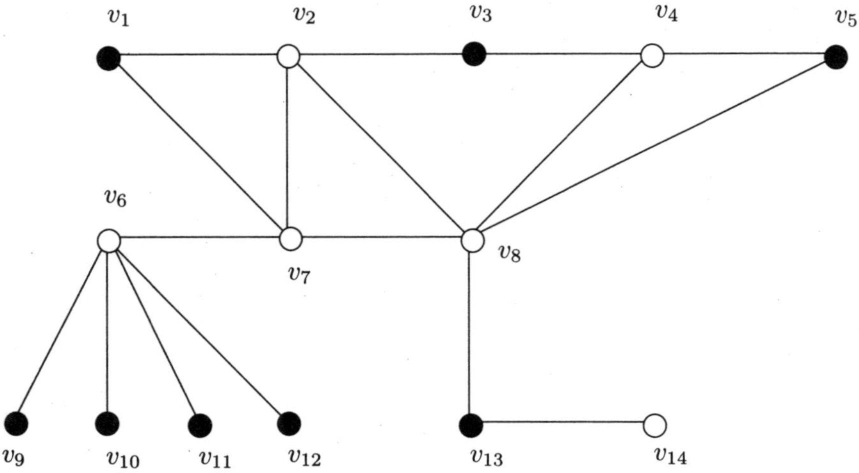

Figure 3.4. *A graph together with a minimum vertex cover (white vertices)*

$p_6 = 0.9$, $p_7 = 0.8$, $p_8 = 0.2$, $p_9 = 0.1$, $p_{10} = 0.1$, $p_{11} = 0.1$, $p_{12} = 0.1$, $p_{13} = 0.5$, $p_{14} = 0.4$.

Given a vertex covering C for G, we can define (at least) the following four functionals:

$$\begin{aligned}
E_1(G,C) &= \sum_{v_i \in V} 1_{\{v_i \in C\}} p_i \\
E_2(G,C) &= \sum_{v_i \in V} 1_{\{v_i \in C : p_i \geq p_0\}} \\
E_3(G,C) &= \sum_{v_i \in V} 1_{\{v_i \in C : p_i \geq p_0\}} p_i \\
E_4(G,C) &= \sum_{v_i \in V} 1_{\{v_i \in C\}}.
\end{aligned}$$

Consider now a minimum vertex cover $C^* = \{v_2, v_4, v_6, v_7, v_8, v_{14}\}$ of G. We then have:

$$\begin{aligned}
E_1(G,C^*) &= 3.2 \\
E_4(G,C^*) &= 6
\end{aligned}$$

Considering $p_0 = 0.5$, we have for $E_2(G, C^*)$ and $E_3(G, C^*)$, respectively:

$$\begin{aligned} E_2(G, C^*) &= 3 \\ E_3(G, C^*) &= 2.3 \end{aligned}$$

Consider next another vertex cover $C_1 = \{v_2, v_4, v_7, v_8, v_9, v_{10}, v_{11}, v_{12}, v_{13}\}$ (the white vertices in Figure 3.4). Then:

$$\begin{aligned} E_1(G, C_1) &= 2.7 \\ E_2(G, C_1) &= 2 \\ E_3(G, C_1) &= 1.4 \\ E_4(G, C_1) &= 9 \end{aligned}$$

For E_2 and E_3, only vertices with probability greater than or equal to p_0 will be taken into account in the computation. Consequently, we can work on the subgraph of G induced by these vertices, the edges not contained in this subgraph being covered by vertices with presence probability smaller than p_0. So, optimizing E_2 or E_3 amounts to working only on the subgraph considered. For $p_0 = 0.5$, for example, the subgraph considered, denoted by G', is the one induced by the vertex-set $\{v_4, v_5, v_6, v_7, v_{13}\}$. Let $C_2 = \{v_5, v_6\}$ be a vertex cover for G'. We have $E_2(G, C_2) = E_2(G', C_2) = 2$ and, consequently, $C_2 = \text{argmin}\{E_2\}$. But C_2 is not optimal for E_3. Indeed, let $C_3 = \{v_4, v_7\}$ be another minimum vertex covering for G'; then $\min_C\{E_3(G, C)\} = E_3(G, C_3) = 1.4 < 1.6 = E_3(G, C_2)$; so, the solution minimizing E_2 does not minimize E_3 and vice versa. Furthermore, let us remark that $\text{argmin}\{E_2\}$ is a minimum size covering for G' and, also, optimizing E_3 on G becomes optimizing E_1 on G'.

In all, with respect to the functionals considered, a solution minimizing one of them does not necessarily minimize the other ones. Rather, all of the four functionals tend to choose vertices of low probability, but the role played by the threshold probability p_0 on the choice of the solutions seems to be important.

3.6. Proofs of the results

3.6.1. *Proof of Theorem 3.3*

The upper bound is obvious since it is nothing else than the expression of Theorem 3.2 for $E(G, C, \mathtt{M2})$.

In order to prove the lower bound, set:

$$A_i = \sum_{\substack{V' \subseteq V \\ v_i \in V'}} \Pr[V'] 1_{\{v_i \notin C_3\}} 1_{\{V' \cap C[\Gamma(v_i)] \neq \emptyset\}} 1_{\{V' \cap \bar{C}[\Gamma(v_i)] = \emptyset\}} \qquad [3.6]$$

(see the last term of [3.2] in Proposition 3.2). Then:

$$\begin{aligned} A_i &\leqslant \sum_{\substack{V' \subseteq V \\ v_i \in V'}} \Pr[V'] 1_{\{V' \cap C[\Gamma(v_i)] \neq \emptyset\}} 1_{\{V' \cap \bar{C}[\Gamma(v_i)] = \emptyset\}} \\ &\leqslant \sum_{\substack{V' \subseteq V \\ v_i \in V'}} \Pr[V'] \left(1 - 1_{\{V' \cap C[\Gamma(v_i)] = \emptyset\}}\right) 1_{\{V' \cap \bar{C}[\Gamma(v_i)] = \emptyset\}} \\ &\leqslant p_i \left(\prod_{v_j \in \bar{C}[\Gamma(v_i)]} (1 - p_j) \right. \\ &\qquad \left. - \prod_{v_j \in \bar{C}[\Gamma(v_i)]} (1 - p_j) \prod_{v_j \in C[\Gamma(v_i)]} (1 - p_j) \right) \\ &\leqslant p_i \left(\prod_{v_j \in \bar{C}[\Gamma(v_i)]} (1 - p_j) - \prod_{v_j \in \Gamma(v_i)} (1 - p_j) \right) \end{aligned}$$

Combining the expression for A_i above with the expression for $E(G, C, \mathtt{M3})$ of Proposition 3.2, we easily get the lower bound claimed.

Let us now suppose that $p_i = p$, $\forall v_i \in V$. Then the expression for $E(G, C, \mathtt{M3})$ becomes:

$$E(G, C, \mathtt{M3}) = p|C| - p \sum_{v_i \in C} (1-p)^{|\Gamma(v_i)|} - \sum_{v_i \in C} A_i$$

with $0 \leqslant A_i \leqslant p((1-p)^{|\bar{C}[\Gamma(v_i)]|} - (1-p)^{|\Gamma(v_i)|})$. Since $A_i \geqslant 0$, we have:

$$E(G, C, \mathtt{M3}) \leqslant p|C| - p \sum_{v_i \in C} (1-p)^{|\Gamma(v_i)|}$$

Using, $\forall i$, $|\Gamma(v_i)| \leqslant \Delta(G)$, we get:

$$E(G, C, \mathtt{M3}) \leqslant p\left(|C| - |C|(1-p)^{\Delta(G)}\right) \qquad [3.7]$$

Note that:

$$p \sum_{v_i \in C} (1-p)^{|\Gamma(v_i)|} = p(1-p)^{\delta_C} \sum_{v_i \in C} (1-p)^{|\Gamma(v_i)|-\delta_C}$$

$$\sum_{v_i \in C} (1-p)^{|\Gamma(v_i)|-\delta_C} \geqslant 1$$

because there exists at least a vertex $v_{i_0} \in C$ with $|\Gamma(v_{i_0})| = \delta_C$. Consequently:

$$E(G, C, \texttt{M3}) \leqslant p\left(|C| - (1-p)^{\delta_C}\right) \qquad [3.8]$$

Combining [3.7] and [3.8], one immediately obtains the upper bound claimed.

For identical vertex-probabilities, the lower bound for $E(G, C, \texttt{M3})$ becomes:

$$p|C|\left(1 - (1-p)^{|\bar{C}[\Gamma(v_i)]|}\right)$$

Since C is minimal, $\bar{C}[\Gamma(v_i)] \neq \emptyset$, i.e., $|\bar{C}[\Gamma(v_i)]| \geqslant 1$. Using the latter inequality in the former one we get $E(G, C, \texttt{M3}) \geqslant p|C|(1-(1-p)) = p^2|C|$, thus proving the lower bound claimed.

Consider now a bipartite graph $B(V_1, V_2, E)$ and one of its color classes, say color class V_1, as an *a priori* solution. Note that $\forall v_i \in V_1$, $\Gamma(v_i) = \bar{V_1}[\Gamma(v_i)] \subseteq V_2$. Consequently, the upper and lower bounds for $E(B, V_1, \texttt{M3})$ coincide.

3.6.2. *On the the bounds obtained in Theorem 3.3*

We now show that there exist instances of PROBABILISTIC MIN VERTEX COVER for which $E(G, C, \texttt{M3})$, computed in Proposition 3.2, can be arbitrarily close to, or arbitrarily far from, the bounds given by Theorem 3.3.

Let $G(V, E)$ be a graph consisting of a clique K_ℓ (on ℓ vertices) and of an independent set S on σ vertices; moreover consider that any vertex of K_ℓ is linked to any vertex of S. Let us suppose that all the vertices of G have the same probability p and consider $C = V(K_\ell)$, the vertex-set of K_ℓ, as an *a priori* solution. Finally set $n = \ell + \sigma$.

For $v_i \in C$ we have $C[\Gamma(v_i)] = V(K_\ell) \setminus \{v_i\}$ and $\bar{C}[\Gamma(v_i)] = S$. Then, [3.2] becomes:

$$\begin{aligned} E(G,C,\mathtt{M3}) &= \sum_{v_i \in C} p - \sum_{v_i \in C} p \prod_{v_j \in \Gamma(v_i)} (1-p) \\ &\quad - \sum_{v_i \in C} \sum_{\substack{V' \subseteq V \\ v_i \in V'}} \Pr[V'] \, 1_{\{v_i \notin C'_3\}} \\ &\quad \times 1_{\{V' \cap (V(K_\ell) \setminus \{v_i\}) \neq \emptyset\}} 1_{\{V' \cap S = \emptyset\}} \end{aligned} \qquad [3.9]$$

Let $V(K_\ell) = \{v_1, v_2, v_3, \ldots, v_\ell\}$ (the degrees of the vertices of $V(K_\ell)$ are all equal) and let $S = \{v_{\ell+1}, v_{\ell+2}, v_{\ell+3}, \ldots, v_n\}$. In step i, strategy M3 tests if vertex v_i can be removed.

Revisit now the third term of [3.9] (the double sum). This term deals with instances $G[V']$ such that $V' \subseteq V(K_\ell) \setminus \{v_i\}$, in other words, instances where all present vertices are elements of $V(K_\ell)$, i.e., they are also part of C. Since $v_i \in V(K_\ell)$, $\Gamma(v_i) \cap V' \subseteq C \cap V'$. Thus, vertex $v_i \in C$ can be removed only if the following two conditions hold:

– v_i is not isolated;

– for $j < i$, $v_j \notin V'$; this is because, in the opposite case, v_j would be removed at step $j < i$, i.e., before v_i; in such a case, since $(v_i, v_j) \in E(K_\ell)$, v_i should not be removed.

Consequently, $1_{\{v_i \notin C'_3\}} = 1_{\{V' \cap \{v_1, v_2, v_3, \ldots, v_{i-1}\} = \emptyset\}}$ and for a fixed index i, [3.6] becomes:

$$\begin{aligned} A_i &= \sum_{\substack{V' \subseteq V \\ v_i \in V'}} \Pr[V'] \, 1_{\{V' \cap \{v_1, v_2, v_3, \ldots, v_{i-1}\} = \emptyset\}} \\ &\quad \times 1_{\{V' \cap V(K_\ell) \setminus \{v_i\} \neq \emptyset\}} 1_{\{V' \cap S = \emptyset\}} \\ &= \sum_{\substack{V' \subseteq V \\ v_i \in V'}} \Pr[V'] \, 1_{\left\{\substack{V' \cap \{v_i\} = \{v_i\} \\ V' \cap \{v_1, v_2, v_3, \ldots, v_{i-1}\} = \emptyset \\ V' \cap \{v_{i+1}, v_{i+2}, v_{i+3}, \ldots, v_\ell\} \neq \emptyset}\right\}} 1_{\{V' \cap S = \emptyset\}} \\ &= \prod_{j=1}^{i-1} (1-p) p \left(1 - \prod_{j=i+1}^{\ell} (1-p)\right) \prod_{j=\ell+1}^{n} (1-p) \\ &= (1-p)^{i-1} p \left(1 - (1-p)^{\ell-(i+1)+1}\right) (1-p)^{n-(\ell+1)+1} \\ &= (1-p)^{i-1} p \left(1 - (1-p)^{\ell-i}\right) (1-p)^{n-\ell} \end{aligned}$$

So we have:

$$\begin{aligned}
\sum_{v_i \in C} A_i &= \sum_{i=1}^{\ell} p\left((1-p)^{n+i-\ell-1} - (1-p)^{n-1}\right) \\
&= p(1-p)^{n-\ell-1} \sum_{i=1}^{\ell} (1-p)^i - p\ell(1-p)^{n-1} \\
&= (1-p)^{n-\ell-1}\left(1-p-(1-p)^{\ell+1}\right) - p\ell(1-p)^{n-1} \\
&= (1-p)^{n-\ell} - (1-p)^n - p\ell(1-p)^{n-1} \qquad [3.10]
\end{aligned}$$

Note that computation of A_is in the expression above is polynomial and, consequently, the whole computation for $E(G, C, \mathtt{M3})$ is also polynomial.

Combining now [3.9] and [3.10], we get:

$$\begin{aligned}
E(G, C, \mathtt{M3}) &= \sum_{i=1}^{\ell} p - \sum_{i=1}^{\ell} p \prod_{v_j \in \Gamma(v_i)} (1-p) \\
&\quad - \left((1-p)^{n-\ell} - (1-p)^n - p\ell(1-p)^{n-1}\right) \\
&= \ell p - \ell p(1-p)^{n-1} - (1-p)^{n-\ell} \\
&\quad + (1-p)^n + p\ell(1-p)^{n-1} \\
&= \ell p - (1-p)^{n-\ell}\left(1-(1-p)^{\ell}\right)
\end{aligned}$$

For facility, set:

$$e(\ell, p, n) \stackrel{\text{def}}{=} \ell p - (1-p)^{n-\ell}\left(1-(1-p)^{\ell}\right) \qquad [3.11]$$

$$b(\ell, p, n) \stackrel{\text{def}}{=} \ell p\left(1-(1-p)^{n-\ell}\right) \qquad [3.12]$$

$$B(\ell, p, n) \stackrel{\text{def}}{=} \ell p\left(1-(1-p)^{n-1}\right) \qquad [3.13]$$

and note that [3.12] and [3.13] are, respectively, the lower and upper bounds of Theorem 3.3 for $E(G, C, \mathtt{M3})$.

We now study the difference $(B-e)(\ell,p,n)$ ([3.11] and [3.13]). We have:

$$(B-e)(\ell,p,n) = (1-p)^{n-\ell} - (1-p)^n - \ell p(1-p)^{n-1} \leqslant 1 \qquad [3.14]$$

Set $x = 1-p$. Then, [3.14] becomes:

$$\begin{aligned}(B-e)(\ell,x,n) &= x^{n-\ell} - x^n - \ell(1-x)x^{n-1} \\ &= x^{n-\ell} + (\ell-1)x^n - \ell x^{n-1} \\ &= x^{n-\ell}\left(1+(\ell-1)x^{\ell} - \ell x^{\ell-1}\right) \\ &\leqslant x^{n-\ell} - x^{n-1} \leqslant 1 \end{aligned} \qquad [3.15]$$

Set $f(x) = 1+(\ell-1)x^{\ell} - \ell x^{\ell-1}$. Then, $f'(x) = \ell(\ell-1)(x^{\ell-1} - x^{\ell-2}) < 0$. Consequently, $f(x)$ is decreasing in $x \in [0,1]$, therefore $f(x) \geqslant f(1) = 0$ and the following holds:

$$0 \leqslant D(\ell,n) \stackrel{\text{def}}{=} (B-e)(\ell,p,n) \qquad [3.16]$$

Revisit [3.15] and set $g(x) = x^{n-\ell} - x^{n-1}$ in [0,1]. First remark that $g(0) = g(1) = 0$. Moreover, $g'(x) = (n-\ell)x^{n-\ell-1} - (n-1)x^{n-2}$ and:

$$g'(x_0) = 0 \iff x_0 = \left(\frac{n-\ell}{n-1}\right)^{\frac{1}{\ell-1}}$$

So, $g(x)$ increases with x, in $[0,x_0)$, while it is decreasing in x in $(x_0,1]$. Hence:

$$(B-e)(\ell,x,n) \leqslant g(x) \leqslant g(x_0) = \left(\frac{n-\ell}{n-1}\right)^{\frac{n-\ell}{\ell-1}} - \left(\frac{n-\ell}{n-1}\right)^{\frac{n-1}{\ell-1}} \qquad [3.17]$$

By [3.16] and [3.17], the following framing holds for $D(\ell,n)$:

$$0 \leqslant D(\ell,n) \stackrel{\text{def}}{=} (B-e)(\ell,p,n) \leqslant \left(\frac{n-\ell}{n-1}\right)^{\frac{n-\ell}{\ell-1}} - \left(\frac{n-\ell}{n-1}\right)^{\frac{n-1}{\ell-1}} \qquad [3.18]$$

Note that the upper bound of [3.18] is quite tight for $D(\ell,n)$. Let us now estimate this bound for several values of ℓ:

– if $\ell = 1$, then $D(\ell, n) = 0$;
– if $1 < \ell \lll n$ (e.g., $\ell = o(n)$), then $\lim_{n\to\infty} D(\ell, n) = 0$;
– if $\ell = \lambda n$, for a fixed $\lambda < 1$, then:

$$\lim_{n\to\infty} D(\lambda n, n) = e^{\frac{1-\lambda}{\lambda}\ln(1-\lambda)} - e^{\frac{\ln(1-\lambda)}{\lambda}}$$

and this limit's value is a fixed positive constant;
– if, finally, $\ell = n - 1$, then $\lim_{n\to\infty} D(n-1, n) = 1$.

Let us now study the difference $(e - b)(\ell, p, n)$. We first have:

$$\begin{aligned}(e-b)(\ell,p,n) &= \ell p(1-p)^{n-\ell} - (1-p)^{n-\ell} + (1-p)^n \\ &\leqslant (\ell-1)\left((1-p)^{n-\ell} - (1-p)^{n-\ell+1}\right)\end{aligned} \quad [3.19]$$

Set $x = 1 - p$. Then, [3.19] becomes:

$$\begin{aligned}(e-b)(\ell,x,n) &= \ell(1-x)x^{n-\ell} - x^{n-\ell} + x^n \\ &= (\ell-1)x^{n-\ell} - \ell x^{n-\ell+1} + x^n \\ &= x^{n-\ell}\left(x^\ell - \ell x + (\ell-1)\right) \\ &\leqslant (\ell-1)\left(x^{n-\ell} - x^{n-\ell+1}\right)\end{aligned} \quad [3.20]$$

Set $h(x) = x^\ell - \ell x + (\ell - 1)$. Then, $h'(x) = \ell(x^{\ell-1} - 1) < 0$, so, $h(x)$ is decreasing in $x \in [0, 1]$ and, consequently, $h(x) \geqslant h(1) = 0$; hence:

$$0 \leqslant d(\ell, n) \stackrel{\text{def}}{=} (e - b)(\ell, p, n) \quad [3.21]$$

Revisit [3.20] and set $\phi(x) = x^{n-\ell} - x^{n-\ell+1}$. Then:

$$\phi'(x) = (n-\ell)x^{n-\ell-1} - (n-\ell+1)x^{n-\ell}$$

and:

$$\phi'(x_0) = 0 \iff x_0 = \frac{n-\ell}{n-\ell+1}$$

So, $\phi(x)$ increases in $[0, x_0)$ and decreases in $(x_0, 1]$. Hence, in [0,1]:

$$\begin{aligned}(e-b)(\ell,x,n) &\leqslant (\ell-1)\left(\frac{n-\ell}{n-\ell+1}\right)^{n-\ell}\left(1-\frac{n-\ell}{n-\ell+1}\right)\\ &\leqslant (\ell-1)e^{(n-\ell)\ln\left(1-\frac{1}{n-\ell+1}\right)}\frac{1}{n-\ell+1}\end{aligned} \qquad [3.22]$$

So, from [3.21] and [3.22], we get the following:

$$0 \leqslant d(\ell,n) \leqslant (\ell-1)e^{(n-\ell)\ln\left(1-\frac{1}{n-\ell+1}\right)}\frac{1}{n-\ell+1} \qquad [3.23]$$

As previously, we estimate bounds in [3.23] for several values of ℓ:
– if $\ell = 1$, then $d(\ell, n) = 0$;
– if $\ell = o(n)$, then $\lim_{n\to\infty} d(\ell, n) = 0$;
– if $\ell = \lambda n$, for a fixed $\lambda < 1$, then:

$$d(\lambda n, n) \leqslant \frac{\lambda n - 1}{(1-\lambda)n+1}e^{n(1-\lambda)\ln\left(1-\frac{1}{(1-\lambda)n+1}\right)} \rightsquigarrow \frac{\lambda}{1-\lambda}e^{-1}$$

and this last value is a fixed constant;
– if, finally, $\ell = n-1$, then $d(n-1, n) \sim n/4$.

It can be seen that for $\ell = o(n)$, $E(G, C, \mathtt{M3})$ tends (for $n \to \infty$) to both $b(\ell, n, p)$, and $B(\ell, n, p)$. This is due to the fact that $o(n)/\lim_{n\to\infty}(B-b)(\ell, n, p) \rightsquigarrow 0$.

In all, we have exhibited instances of PROBABILISTIC VERTEX COVER3 and *a priori* solutions for these instances for which, for $n \to \infty$, the distance of $E(G, C, \mathtt{M3})$ from the bounds given in Theorem 3.3 can take an infinity of values being either arbitrarily close to or arbitrarily far from them. The way $E(G, C, \mathtt{M3})$ has been computed in this chapter (Proposition 3.2 of section 3.4.1) is not the only one of its kind. There would be many other ones (even if we do not see which). But for our way, it seems very unlikely that the bounds computed could be generally improved in order that the new bounds hold for any graph. So, if we continue adopting Proposition 3.2 for $E(G, C, \mathtt{M3})$, we think that it is very interesting to search for bounds of $E(G, C, \mathtt{M3})$ on particular classes of graphs and, perhaps, under particular systems of vertex-probabilities.

Chapter 4

The Probabilistic Longest Path

In this chapter, we study a restrictive but very frequently handled version of PROBABILISTIC LONGEST PATH, the one defined on graphs that are directed acyclic and transitive. The general longest path problem, denoted by LONGEST PATH in what follows, is defined as follows: consider a directed graph $G(V, A)$ and two fixed vertices s and t in V; in LONGEST PATH the objective is to determine a maximum-size directed path from s to t.

If the measure of the problem is expressed by means of vertices (the path's length as number of vertices, or sum of the weights of the vertices on the path), we have the versions of LONGEST PATH which in the sequel we call VERTEX LONGEST PATH, or VERTEX WEIGHTED LONGEST PATH, while if the measure of the problem is expressed by means of arcs (the path's length as number of arcs, or sum of the weights of the arcs on the path) we have the variants called ARC LONGEST PATH, or ARC WEIGHTED LONGEST PATH, in what follows. In the deterministic framework, the unweighted problems are equivalent in the sense that, for a given graph, the solution value of VERTEX LONGEST PATH differs from the solution value of ARC LONGEST PATH by 1.

The probabilistic versions of all of VERTEX LONGEST PATH, ARC LONGEST PATH, VERTEX WEIGHTED LONGEST PATH and ARC WEIGHTED LONGEST PATH, PROBABILISTIC VERTEX LONGEST PATH, PROBABILISTIC ARC LONGEST PATH, PROBABILISTIC VERTEX WEIGHTED LONGEST PATH and PROBABILISTIC ARC WEIGHTED LONGEST PATH, respectively, are built considering presence probabilities associated with the vertices of the input-graph (in what follows, we will denote, as is usual, by p_i the presence probability of vertex $v_i \in V$). The only restriction admitted on the form of these probabilities is that the probabilities of s and t (the initial and terminal endpoints of the path) are equal to 1 (i.e., s and t are always present). The objective for the probabilistic versions dealt is, as we have already mentioned, given a modification strategy, to compute the functionals associated with, and then to determine the *a priori* solution maximizing them.

Given a graph $G(V \cup \{s,t\}, A)$, we set $L_V = (v_0, v_1, \ldots, v_k, v_{k+1})$ (list of vertices) and $L_A = (a_0, a_1, \ldots, a_k)$ (list of arcs), where $v_0 = s$, $v_{k+1} = t$, $a_i = (v_i, v_{i+1})$, $i = 0, \ldots, k$. In other words, L_V and L_A are the lists of vertices and arcs representing the *a priori* solutions (paths) for the variants (simple and weighted) of PROBABILISTIC VERTEX LONGEST PATH and PROBABILISTIC ARC LONGEST PATH, respectively. For simplicity, an arc (v_i, v_j) will be called transitive if there exists a path from v_i to v_j using at least two arcs. By analogy if such a path does not exist, then (v_i, v_j) will be called non-transitive. Given a set $V' \subseteq V$ containing vertices s and t, we denote by $A(V')$ the arc-set of $G[V']$. Recall that the "occurrence" probability of $G[V']$ is $\Pr[V'] = \prod_{v_i \in V'} p_i \prod_{v_i \in V \setminus V'} (1 - p_i)$. Also, instead of $E(G, S, \mathtt{M})$, we use notation $E(G, \vec{c}, S, \mathtt{M})$ in order to denote the functional for PROBABILISTIC VERTEX WEIGHTED LONGEST PATH associated with *a priori* solution S and modification strategy M in a vertex-weighted graph with vertex-weight vector $\vec{c}$; analogously, for PROBABILISTIC ARC WEIGHTED LONGEST PATH we will use notation $E(G, \vec{d}, S, \mathtt{M})$, where $\vec{d}$ is the vector of the arc-weights. Also for a vertex-set V' (resp., arc-set A'), we denote by $c(V')$ (resp., $d(A')$) the total weight of V' (resp., A').

4.1. Probabilistic longest path in terms of vertices

The modification strategy MV adopted for both PROBABILISTIC VERTEX LONGEST PATH and PROBABILISTIC VERTEX WEIGHTED LONGEST PATH is strategy MS (see Chapter 1) adapted to fit PROBABILISTIC VERTEX LONGEST PATH. It consists, given an *a priori* solution L_V and a present set of vertices V', of dropping the absent vertices out of L_V. It is easy to see that, because of the transitivity of G, the solution L'_V thus obtained is feasible for $G[V']$.

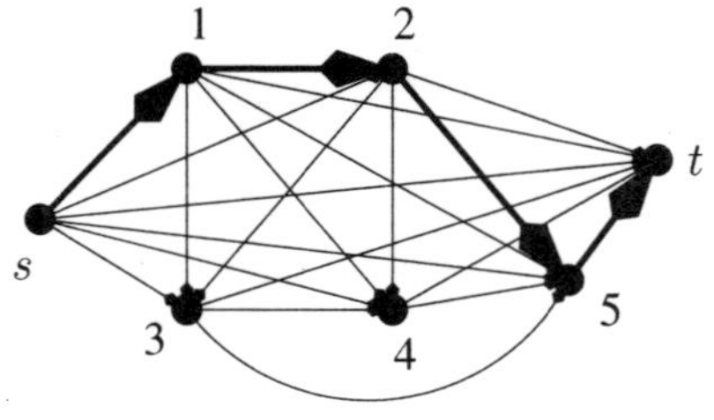

Figure 4.1. *A graph G input of* PROBABILISTIC VERTEX LONGEST PATH, *and an a priori solution* $\{s, 1, 2, 5, t\}$

EXAMPLE 4.1.– Consider the graph of Figure 4.1, set $L_V = \{s, 1, 2, 5, t\}$ (some *a priori* solution), and assume that vertices 1 and 4 are absent. Then application of MV produces the path $L'_V = \{s, 2, 5, t\}$ (Figure 4.2).

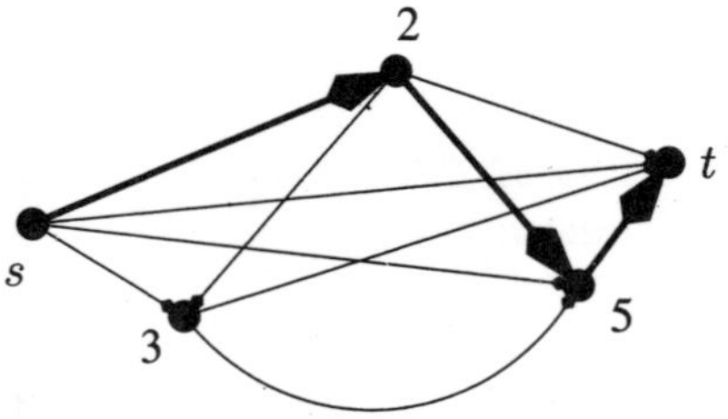

Figure 4.2. *Application of strategy MV on $G[V']$, with an a priori solution $\{s, 1, 2, 5, t\}$; path produced: $\{s, 2, 5, t\}$*

For VERTEX WEIGHTED LONGEST PATH, the value of L_V is $\sum_{v_i \in L_V} c(v_i)$. For PROBABILISTIC VERTEX WEIGHTED LONGEST PATH, MV provides for $G[V']$ a solution L'_V of value:

$$c(L'_V) = 2 + \sum_{v_i \in L'_V \cap V} c(v_i)\, 1_{\{v_i \in V'\}} \quad [4.1]$$

The functional associated with PROBABILISTIC VERTEX WEIGHTED LONGEST PATH is:

$$E(G, \vec{c}, L_V, \mathtt{MV}) = \sum_{V' \subseteq V} \Pr[V']\, c(L'_V) \quad [4.2]$$

In what follows, we compute the functional in a more explicit way, we give the complexity of such a computation and we determine the *a priori* solution maximizing $E(G, \vec{c}, L_V, \mathtt{MV})$. The following basic result holds (its proof can be found in Section 4.4.1.

THEOREM 4.1.– *For* PROBABILISTIC VERTEX WEIGHTED LONGEST PATH, *the functional $E(G, \vec{c}, L_V, MV)$ is given by:*

$$E(G, \vec{c}, L_V, MV) = 2 + \sum_{v_i \in L_V \cap V} p_i c(v_i)$$

It is computed in $O(n)$. The a priori solution L_V maximizing $E(G, \vec{c}, L_V, MV)$ is an optimal solution of VERTEX WEIGHTED LONGEST PATH *in G (where the cost for* *a vertex v_i is the quantity $p_i c(v_i))$ and is computed in polynomial time. Finally, if $p_i = p$, $v_i \in V$, then:*

$$E(G, \vec{c}, L_V, MV) = 2 + p \sum_{v_i \in L_V \cap V} c(v_i)$$

For the case of PROBABILISTIC VERTEX LONGEST PATH strategy MV provides for $G[V']$ a solution L'_V of cardinality:

$$|L'_V| = 2 + \sum_{v_i \in L'_V \cap V} 1_{\{v_i \in V'\}} = k + 2 - \sum_{v_i \in L'_V \cap V} 1_{\{v_i \notin V'\}} \qquad [4.3]$$

Using [4.3] and what has been discussed above, the following corollary summarizes the case for PROBABILISTIC VERTEX LONGEST PATH.

COROLLARY 4.1.– $E(G, L_V, \texttt{MV}) = 2 + \sum_{v_i \in L_V \cap V} p_i$ and is computed in $O(n)$. The *a priori* solution L_V maximizing $E(G, L_V, \texttt{MV})$ is an optimal solution of VERTEX WEIGHTED LONGEST PATH in G (where the vertex-costs are the vertex-probabilities) and is computed in polynomial time. Finally, if $p_i = p$, $v_i \in V$, then $E(G, L_V, \texttt{MV}) = 2 + pk$.

4.2. Probabilistic longest path in terms of arcs

Let $L_A = (a_0, a_1, \ldots, a_k)$, where $a_i = (v_i, v_{i+1})$, $v_0 = s$ and $v_{k+1} = t$ be a path of length (number of arcs) $k + 1$ and let $G[V'](V', A(V'))$, $\{s, t\} \subset V'$ be the present subgraph of G. Then, the following modification strategy MA is used for obtaining a PROBABILISTIC ARC WEIGHTED LONGEST PATH- (or PROBABILISTIC ARC LONGEST PATH-) solution for $G[V']$:

– set $L'_A = L_A \cap A(V')$;

– if $a_0 \notin L'_A$ and a_i is the first arc of L'_A, then set $L'_A = L'_A \cup \{(v_0, v_i)\}$;

– if (v_i, v_{i+1}) and (v_j, v_{j+1}), $j > i + 1$, are consecutive arcs in L'_A, then set $L'_A = L'_A \cup \{(v_{i+1}, v_j)\}$;

– if $a_k \notin L'_A$ and a_l is the last arc of L', then set $L'_A = L'_A \cup \{(v_{l+1}, t)\}$;

– output L'_A.

It is easy to see that, thanks to the transitivity of G, solution L'_A, constructed by strategy MA is feasible.

EXAMPLE 4.2.– Consider the graph of Figure 4.1, assume an *a priori* solution given by set $L_A = \{(s, 1), (1, 2), (2, 5), (5, t)\}$, and assume that vertices 1 and 4 are absent. Then application of MA produces the path $L'_A = \{(s, 2), (2, 5), (5, t)$ (Figure 4.2).

THEOREM 4.2.– *The functional $E(G, \vec{d}, L_A, \mathtt{MA})$ associated with* PROBABILISTIC ARC WEIGHTED LONGEST PATH *and with modification strategy* MA *is given by:*

$$E\left(G, \vec{d}, L_A, \mathtt{MA}\right) = \sum_{i=0}^{k} p_i p_{i+1} d\left(v_i, v_{i+1}\right) + \sum_{i=0}^{k-1} \sum_{j=i+2}^{k+1} p_i p_j \left(\prod_{l=i+1}^{j-1} \left(1 - p_l\right)\right) d\left(v_i, v_j\right)$$

and is computed in $O(n^3)$.

The proof of Theorem 4.2 is in section 4.4.2.

In a similar way than the one followed in Theorem 4.2, we can obtain several equivalent expressions for $E(G, \vec{d}, L_A, \mathtt{MA})$. The most interesting is the one given in Proposition 4.1.

PROPOSITION 4.1.– *Let us suppose that for $i' < i$, $\prod_{l=i}^{i'}(1 - p_l) = 1$. Then:*

$$E\left(G, \vec{d}, L_A, \mathtt{MA}\right) = \sum_{i=1}^{k-1} \sum_{j=i+1}^{k} p_i p_j \left(\prod_{l=i+1}^{j-1} \left(1 - p_l\right)\right) d\left(v_i, v_j\right) + \prod_{l=1}^{k} \left(1 - p_l\right) d(s,t) + \sum_{j=1}^{k} p_j \left(\prod_{l=1}^{j-1} \left(1 - p_l\right)\right) d\left(s, v_j\right) + \sum_{i=1}^{k} p_i \left(\prod_{l=i+1}^{k} \left(1 - p_l\right)\right) d\left(v_i, t\right)$$

If, for $v_i \in V$, $p_i = p$, then:

$$E\left(G, \vec{d}, L_A, \mathtt{MA}\right) = (1-p)^k d(s,t) + p \sum_{i=1}^{k} \left((1-p)^{i-1} d\left(s, v_i\right) + (1-p)^{k-i} d\left(v_i, t\right)\right) + p^2 \sum_{i=1}^{k-1} \sum_{j=i+1}^{k} (1-p)^{j-i-1} d\left(v_i, v_j\right)$$

As one can see from the expressions for $E(G, \vec{d}, L_A, \texttt{MA})$ given in both Theorem 4.2 and Proposition 4.1, if one tries to express PROBABILISTIC ARC WEIGHTED LONGEST PATH, under strategy MA, as a kind of arc-weighted deterministic longest path problem, then the weights that have to be assumed depend on the *a priori* solution itself. So, under these two functional-expressions, it is difficult to conclude on a precise characterization of the *a priori* solution maximizing them.

4.2.1. *An interesting algebraic expression*

In this section we deal with a simplified version of PROBABILISTIC ARC WEIGHTED LONGEST PATH, namely the one of PROBABILISTIC ARC LONGEST PATH (where all arcs of G have equal distances). For this restrictive version, the following result can be proved.

THEOREM 4.3.– $E(G, L_A, \texttt{MA}) = 1 + \sum_{i=1}^{k} p_i$ *and is computable in* $O(n)$. *The a priori solution maximizing it is computed in polynomial time. If all the vertices of G have the same presence-probability p, then* $E(G, L_A, \texttt{MA}) = 1 + pk$.

Proof. We first prove that in the case of unit (or equal) arc-distances, for every subinstance V' of G, the value of the optimal solution in terms of vertices equals the value of the optimal solution in terms of arcs plus 1. For this, it suffices to prove that in every subgraph $G[V']$ of G, solutions computed by procedures MV and MA are identical.

Consider an optimal *a priori* solution μ for G and a subgraph $G[V']$ of G. Also let $\mu_1(V') = (s, \ldots, v_i, v_j, \ldots, t)$ and $\mu_2(V') = (s, \ldots, v_i, v_k, \ldots, t)$ be the solutions computed by strategies MV and MA in $G[V']$, respectively. Suppose that $\mu_1(V')$ and $\mu_2(V')$ are identical from s to v_i and that $j < k$. Finally, suppose that both solutions are expressed in terms of vertices and recall that the vertex-indices in both lists are sorted in increasing order (see notations in the beginning of the chapter). Obviously, the vertices of both $\mu_1(V')$ and $\mu_2(V')$ also appear in μ. Following the hypotheses just made, v_j is present in $G[V']$ and, moreover, arc (v_i, v_k) is part of the PROBABILISTIC ARC LONGEST PATH-solution computed by MA. So we have to examine two possible cases:

– $(v_i, v_k) \in \mu$ when μ is seen as PROBABILISTIC ARC LONGEST PATH solution; this is impossible since there exists at least v_j which appears in μ between v_i and v_k;

– $(v_i, v_k) \notin \mu$; consequently there exists a subpath $\mu' \subset \mu$ from v_i to v_k and any vertex v_l with $i < l < k$ is absent from $G[V']$; this is impossible since v_j with $i < j < k$ is supposed present in $G[V']$.

Hence, $\mu_1(V')$ and $\mu_2(V')$ are identical and, consequently, for any V', $|L'_A| = |L'_V| - 1 = |\mu_1(V')| - 1$; therefore:

$$\begin{aligned} E\left(G, L_A, \mathtt{MA}\right) &= \sum_{V' \subseteq V} \Pr\left[V'\right] |L'_A| = \sum_{V' \subseteq V} \Pr\left[V'\right] \left(|L'_V| - 1\right) \\ &= E\left(G, L_V, \mathtt{MV}\right) - 1 = 1 + \sum_{i=1}^{k} p_i \end{aligned} \quad [4.4]$$

where the last equality holds by Theorem 4.1.

Consequently, $E(G, L_A, \mathtt{MA})$ can be computed with worst-case time complexity linear in n.

Since, from [4.4], $E(G, L_A, \mathtt{MA}) = E(G, L_V, \mathtt{MV}) - 1$, the same *a priori* solution (an optimal solution for VERTEX WEIGHTED LONGEST PATH on G with its vertices weighted by their probabilities) simultaneously maximizes both $E(G, L_V, \mathtt{MV})$ and $E(G, L_A, \mathtt{MA})$; therefore, by Corollary 4.1, determining the *a priori* solution maximizing $E(G, L_A, \mathtt{MA})$ is performed in polynomial time.

Finally, by substituting, in [4.4], p_i by p, $v_i \in V$ (for the case of identical vertex-probabilities), we obtain $E(G, L_A, \mathtt{MA}) = 1 + pk$. ∎

The proof of Theorem 4.3 is established using combinatorial arguments. We show in section 4.4.3, that Theorem 4.3 can be established also by purely algebraic arguments. This proof has its own (algebraic) interest going beyond probability- or graph-theory. Indeed, as can be seen in section 4.4.3, the last argument of this proof (expression [4.15]) establishes the following corollary holding for all tuples of $n + 2$ numbers of the form $(1, x_1, \ldots, x_n, 1)$ with $x_i \leqslant 1$, $1 \leqslant i \leqslant n$.

COROLLARY 4.2.– Consider $n + 2$ numbers $x_i \leqslant 1$, $i = 0, \ldots, n + 1$ with $x_0 = x_{n+1} = 1$. Then:

$$\sum_{i=0}^{n} x_i x_{i+1} + \sum_{i=0}^{n-1} \sum_{j=i+2}^{n+1} x_i x_j \left(\prod_{l=i+1}^{j-1} (1 - x_l) \right) = 1 + \sum_{i=1}^{n} x_i$$

4.2.2. *Metric* PROBABILISTIC ARC WEIGHTED LONGEST PATH

We study in what follows the complexity of the metric PROBABILISTIC ARC WEIGHTED LONGEST PATH. For this, we shall first give a characterization of the *a priori* solution maximizing $E(G, \vec{d}, L_A, \texttt{MA})$ (i.e., of the optimal *a priori* solution). For this, we require the following definition introducing a notion of path domination.

DEFINITION 4.1.– *A path μ' is said* dominated *if there exists a path μ such that: $E(G, \vec{d}, \mu, \textit{MA}) \geqslant E(G, \vec{d}, \mu', \textit{MA})$.*

We prove now that a path using transitive arcs is always a dominated one. For this, we first use the two following Lemmata 4.1 and 4.2 (the proofs of which can be found in sections 4.4.4 and 4.4.5, respectively).

LEMMA 4.1.– *Consider a list $(x_1, \ldots, x_n)$ of numbers and let i and m be two indices such that $1 \leqslant i \leqslant m \leqslant n$. Then:*

$$\sum_{j=i}^{m} x_j \prod_{l=i}^{j-1} (1 - x_l) = 1 - \prod_{l=i}^{m} (1 - x_l)$$

LEMMA 4.2.– *Consider a list $(x_0, \ldots, x_n)$ with $0 \leqslant x_i \leqslant 1$, $i = 1, \ldots, n$. Then:*

$$\sum_{i=0}^{n-2} x_i \prod_{j=i+1}^{n} (1 - x_j) \leqslant (1 - x_{n-1})(1 - x_n)$$

THEOREM 4.4.– *Consider two paths:*

$$\begin{aligned} \mu &= (a_0, a_1, \ldots, (v_{i_0-1}, v_{i_0}), (v_{i_0}, v_{i_1}), (v_{i_1}, v_{i_2}), (v_{i_2}, v_{i_2+1}), \ldots, a_k) \\ \mu' &= (a_0, a_1, \ldots, (v_{i_0-1}, v_{i_0}), (v_{i_0}, v_{i_2}), (v_{i_2}, v_{i_2+1}), \ldots, a_k) \end{aligned}$$

In other words, both paths are from s to t and μ' contains the same vertices as μ except v_{i_1}. Then, under the modification strategy induced by procedure MA:

$$E\left(G, \vec{d}, \mu, \textit{MA}\right) \geqslant E\left(G, \vec{d}, \mu', \textit{MA}\right)$$

in other words, μ' is dominated.

The proof of Theorem 4.4 can be found in section 4.4.6.

An immediate consequence of Theorem 4.4 is that *every path using at least one transitive arc is dominated.* Obviously, the optimal PROBABILISTIC ARC WEIGHTED LONGEST PATH-solution (the *a priori* solution maximizing $E(G, \vec{d}, L_A, \texttt{MA})$) cannot be dominated; so, *this solution corresponds to a path containing only non-transitive arcs.*

LEMMA 4.3.– *Let us denote by G' the undirected version of G. Then:*

– every path from s to t in G corresponds to a clique in G';

– to every maximal (for the inclusion) clique K' of G' containing s and t corresponds a unique path from s to t in G; this path includes all the vertices of K' and uses only non-transitive arcs.

Proof. Let μ be a path from s to t in G. Since G is transitive, for every pair (v_i, v_j) of vertices in μ, there exists an arc (v_i, v_j). So the vertex-set of μ induces in G a completely connected subgraph, i.e, a clique, thus concluding the proof of the first item.

In order to prove the second item, we first notice that s and t belong to at least one maximal clique of G'. In fact, since G is transitive and s is a source of G, it is connected to every other vertex of G. On the other hand, if we consider a clique K' of G' containing at least one neighbor of t (this neighbor is, really, a predecessor of t in G), then, by transitivity, t is linked to every other vertex of K'.

Let us now consider a maximal clique K' in G' and denote by $V(K')$ its vertex-set. It is well-known that the directed version K of K' also being acyclic, it contains one source, s in this case. Moreover, it is easy to see that s is the unique source because, in the opposite case, K' would not be a clique (because there would exist at least two sources in K and, being sources, these would be not mutually linked).

With the same arguments, $V(K') \setminus \{s\}$ induces a clique of G' and its directed version contains a unique source v_0. This vertex is a successor of s and arc (s, v_0) is, obviously, non-transitive.

So, by successive reasoning on the subcliques of K' (resulting from the removal of the unique source of the precedent step), we bring to the fore a path from s to t in G, including all the vertices of K' and using exclusively non-transitive arcs, thus concluding the proof of the second item and of the lemma. ∎

The following Theorem 4.5 results from an immediate simultaneous combination of Theorem 4.2 and Lemma 4.3.

THEOREM 4.5.– *Let G' be the undirected version of G; moreover, given an edge-weighted graph, let us define the edge-weight of a clique as the sum of the weights of its edges. Then, the optimal* PROBABILISTIC ARC WEIGHTED LONGEST PATH*-solution in G corresponds to a maximal (for the inclusion in terms of vertices) maximum edge-weight clique K' of G' where the weights on the edges of K' are computed as follows:*

– if edge (v_i, v_j) corresponds to a non-transitive arc (v_i, v_j) of A, then the weight $w(v_i, v_j)$ is:

$$w(v_i, v_j) = d(v_i, v_j)\, p_i p_j$$

– *if edge* (v_i, v_j) *corresponds to a transitive arc* (v_i, v_j) *of* A, *then:*

$$w(v_i, v_j) = d(v_i, v_j) p_i p_j \prod_{l \in \mu[v_i, v_j]} (1 - p_l)$$

where μ *is the path induced by* K' *(following the second item of Lemma 4.3) and* $\mu[v_i, v_j]$ *is the set of vertices of* μ *lying between* v_i *and* v_j.

Let us recall that a *comparability graph* is a graph, the edges of which can be oriented in such a way that the resulting directed graph is acyclic and transitive (see Appendix A). Obviously, the graph G' of Theorem 4.5 is a comparability graph.

In [CAI 92] (see also [BJO 97]), an $O(\sum_{K \in \mathcal{K}} V(K))$ algorithm is presented generating all maximal cliques in comparability graphs, where $\mathcal{K}$ draws the set of maximal cliques of G'. This result can be immediately used to devise the following algorithm, denoted by METRIC for metric PROBABILISTIC ARC WEIGHTED LONGEST PATH:

– construct G';

– run the algorithm of [CAI 92] in G';

– for any generated clique K':

1) apply Lemma 4.3 to extract the induced path μ;
2) weight the edges of K' by the weights suggested by Theorem 4.5;
3) compute the edge-weight of K';

– let K^* be the maximum-weight clique computed by Steps 1 to 3;

– output the induced path μ^* of K^* as the PROBABILISTIC ARC WEIGHTED LONGEST PATH-solution.

By Theorem 4.5, algorithm METRIC computes the *a priori* solution maximizing the functional of the metric version of PROBABILISTIC ARC WEIGHTED LONGEST PATH. Consequently, the following theorem concludes this section.

THEOREM 4.6.– *Metric* PROBABILISTIC ARC WEIGHTED LONGEST PATH *is optimally solved by algorithm METRIC in* $O(\sum_{K \in \mathcal{K}} V(K))$.

The characterization of the optimal *a priori* solution for metric PROBABILISTIC ARC WEIGHTED LONGEST PATH provided by Theorem 4.5 does not apply for general PROBABILISTIC ARC WEIGHTED LONGEST PATH. Indeed, consider the graph of Figure 4.3 where all vertices have the same probability p except vertices s and t which are of probability 1. All the arcs except (a, c) have weight 1; the weight of (a, c) is M for some large $M \in \mathbb{R}$. Consider also paths $\mu = (sa, ab, bc, ct)$ and $\mu' = (sa, ac, ct)$. Expectations of μ and μ' are, respectively:

$$E\left(G, \vec{d}, \mu, \mathtt{MA}\right) = p + p^2 + p^2 + p + p(1-p) + p(1-p)^2$$

$$+ (1-p)^3 + p(1-p)^2 + p(1-p) + p^2(1-p)M$$

$$E\left(G, \vec{d}, \mu', \mathtt{MA}\right) = p + p^2M + p + p(1-p) + (1-p)^2 + p(1-p)$$

Their difference is:

$$E\left(G, \vec{d}, \mu, \mathtt{MA}\right) - E\left(G, \vec{d}, \mu', \mathtt{MA}\right) = 2p^2 + 2p(1-p)^2 + (1-p)^3$$
$$+ p^2(1-p)M - p^2M - (1-p)^2$$
$$= (1-M)p^3 + p$$

The above difference is negative for $M \geqslant (1/p^2) + 1$; in other words, choosing M arbitrarily large, μ' although containing a transitive arc, is longer than μ.

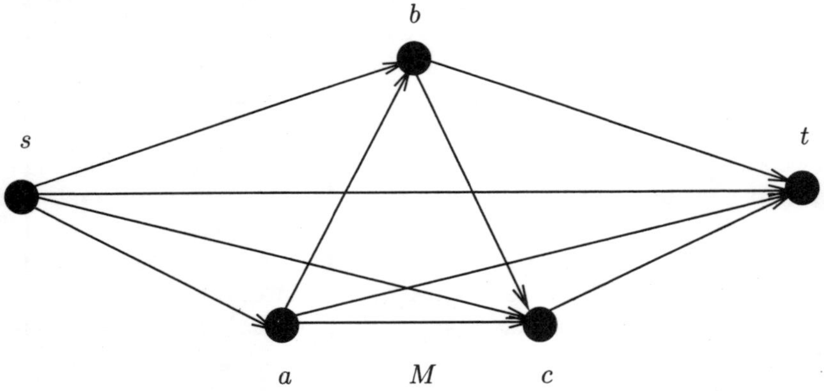

Figure 4.3. *A counter-example for general* PROBABILISTIC ARC WEIGHTED LONGEST PATH

4.3. Why the strategies used are pertinent

The modification strategies MV and MA used in this chapter are quite natural for the problems covered here. Given an *a priori* solution L, they generally consist first of removing absent vertices (disconnecting L) and then in using appropriate arcs in order to reconnect the several surviving pieces of L. These arcs always exist due to the fact that the input-graph has been assumed to be transitive. In fact, as one can see, they are the simplest modification strategies one can invent.

There exist several reasons rendering use of such strategies pertinent for the versions of probabilistic LONGEST PATH we have dealt with in this chapter. These strategies model the probabilistic optimization context very well. Here, we are initially given a "potential" graph and in the sequel we have considered that objects (vertices) disappear at the last moment, too late to react by applying more complicated strategies. Constructing more complicated modification strategies implies that we are able to know the subinstance which will be actually optimized sufficiently early in order to have the time to devise such a strategy. But if one has the time to do this, then it is very likely that one has the time to solve the deterministic problem entirely in the present subinstance (recall that the version of the deterministic longest path we deal with is polynomial), thus obtaining, in polynomial time, an optimal solution.

Some other reasons can also justify the study of strategies MV and MA. In [GAB 97], it is shown that a short-term satellite shot planning problem (with planning-horizon of one day) can be modelled in terms of LONGEST PATH. But uncertainty due to meteorological phenomena is not taken into account in the modelling of [GAB 97]; this can be done by associating probabilities with the vertex-set of the graph-instance of LONGEST PATH and a probabilistic longest path problem has then to be solved. The planning performed under the meteorological uncertainty (of the next day) consists of finding an *a priori* solution (maximizing the functional) and once the shots planned have been taken, the inexploitable shots are removed; this is strategy MV exactly.

The second motivation is more theoretical. As we have discussed in the beginning of the chapter, VERTEX LONGEST PATH and ARC LONGEST PATH are equivalent in the sense that an optimal solution of the former is identical to an optimal solution of the latter and their values differ by 1. Hence VERTEX LONGEST PATH and ARC LONGEST PATH are also of identical complexity. We have seen that strategies MV and MA, of course under different data-representation, provide identical solutions for the corresponding probabilistic versions. Our purpose studying these strategies was also to confirm this fact.

4.4. Proofs of the results

4.4.1. *Proof of Theorem 4.1*

Starting from the expression for $c(L'_V)$ given in (4.1), and since s and t are always present, we get:

$$c\left(L'_V\right) = m\left(G\left[V'\right], L_V\left(V', \texttt{MV}\right)\right) = 2 + \sum_{i=1}^{k} c\left(v_i\right) 1_{\{v_i \in V'\}} \qquad [4.5]$$

A combination of [4.2] and [4.5] leads to:

$$\begin{aligned} E\left(G,\vec{c},L_V,\mathtt{MV}\right) &= \sum_{V'\subseteq V} \Pr\left[V'\right] c\left(L'_V\right) \\ &= \sum_{V'\subseteq V} \Pr\left[V'\right] \left(2+\sum_{i=1}^{k} c\left(v_i\right) 1_{\{v_i\in V'\}}\right) \\ &= 2\sum_{V'\subseteq V} \Pr\left[V'\right] + \sum_{i=1}^{k}\sum_{V'\subseteq V} \Pr\left[V'\right] c\left(v_i\right) 1_{\{v_i\in V'\}} \\ &= 2+\sum_{i=1}^{k} p_i c\left(v_i\right) \end{aligned} \qquad [4.6]$$

which proves the first result of the theorem.

The expression for $E(G, \vec{c}, L_V, \mathtt{MV})$, in the case where all the vertices are of probability p, is obtained by simply substituting p for p_i, $v_i \in V'$. Finally, it is immediately observed that the computation of $E(G, \vec{c}, L_V, \mathtt{MV})$ needs only at most n multiplications (the terms $p_i c(v_i)$) and the addition of the n rational numbers, results of the multiplications. So, the linear complexity of such a computation is concluded.

As one can see from [4.6], the maximum value for $E(G, \vec{c}, L_V, \mathtt{MV})$ is:

$$E^*_{\mathrm{WV}} = 2 + \max_{L_V}\left\{\sum_{v_i\in L_V\cap V} p_i c\left(v_i\right)\right\}$$

and the *a priori* solution L^*_V corresponding to E^*_{WV} is a kind of vertex-weighted longest path in G where the weights on the vertices of G are, this time, the products $p_i c(v_i)$; these quantities may be rational numbers.

In what follows, we shall prove that finding an optimal solution for VERTEX WEIGHTED LONGEST PATH in directed acyclic graphs, when the weights are rational, is polynomial.

It is well-known ([GAR 79]) that ARC WEIGHTED LONGEST PATH is polynomial on directed acyclic graphs. To start with let us prove that even if the weights on the arcs are rational, ARC WEIGHTED LONGEST PATH remains polynomial. To do it, we transform every arc-distance $d_i = p_i/q_i$, $q_i \in \mathbb{N}$, $i = 1, \ldots, m$ into $d'_i = Qd_i$ where Q is, for example, the lcm of the q_is. It is then easy to see that the longest paths in the modified graph remain identical to the ones in G.

Finally, let us show how one can transform VERTEX WEIGHTED LONGEST PATH into ARC WEIGHTED LONGEST PATH in such a way that the corresponding solutions are the same.

Consider a graph $G(V \cup \{s,t\}, E, \vec{c}_n)$, instance of VERTEX WEIGHTED LONGEST PATH. We construct the graph $G'(V \cup \{s,t\}, E, \vec{d}_m)$, instance of ARC WEIGHTED LONGEST PATH, where for an arc (v_i, v_j), $d(v_i, v_j) = c(v_i)$.

As we have already mentioned, ARC WEIGHTED LONGEST PATH from s to t is polynomial in G'; let $\mu' = (a_0, a_1, \dots, a_k)$ be a longest path from s to t in G', with $a_0 = (s, v_{i_1})$, $a_j = (v_{i_j}, v_{i_{j+1}})$, $a_k = (v_{i_k}, t)$, $v_{i_j} \in V$, $j = 1, \dots, k-1$. We shall prove that the path $\mu = (s, v_{i_1}, \dots, v_{i_j}, v_{i_{j+1}}, \dots, v_{i_k}, t)$ is an optimal solution for VERTEX WEIGHTED LONGEST PATH.

Let us suppose that there exists another path:

$$\hat{\mu} = (s, \hat{v}_{i_1}, \dots, \hat{v}_{i_{j-1}}, \hat{v}_{i_j}, \dots, \hat{v}_{i_k}, t)$$

with $\hat{\mu} \neq \mu$, longer than μ in G. Then, $\sum_{\hat{v}_{i_j} \in \hat{\mu}} c(\hat{v}_{i_j}) > \sum_{v_{i_j} \in \mu} c(v_{i_j})$. On the other hand, since, by the construction of G', $c(v_{i_j}) = d(v_{i_j}, v_{i_{j+1}}) = d(a_j)$, if we consider the path $\hat{\mu} = (\hat{u}_0, \hat{u}_1, \dots, \hat{u}_k)$ with $\hat{u}_0 = (s, \hat{v}_{i_1})$, $\hat{u}_j = (\hat{v}_{i_j}, \hat{v}_{i_{j+1}})$, $\hat{u}_k = (\hat{v}_{i_k}, t)$, we get $\sum_{\hat{u}_j \in \hat{\mu}} d(\hat{u}_j) > \sum_{a_j \in \mu'} d(a_j)$, which contradicts the optimality of μ' for ARC WEIGHTED LONGEST PATH. Consequently, VERTEX WEIGHTED LONGEST PATH is polynomially solved.

The above discussion leads to the conclusion that solving VERTEX WEIGHTED LONGEST PATH, even when the vertex-costs are rational, is polynomial; therefore, by the expression of $E(G, \vec{c}, L_V, \mathtt{MV})$, determining the *a priori* solution maximizing it is also polynomial and the proof of the theorem is complete.

4.4.2. *Proof of Theorem 4.2*

Let $V(L_A) = (s, v_1, \dots, v_k, t)$ be the list of vertices associated with the path L_A. Then, $G[V(L_A)](V(L_A), A(L_A))$, the subgraph of G induced by $V(L_A)$ is a transitive directed acyclic graph with:

$$A(L_A) = \bigcup_{i=0}^{k} \bigcup_{j=i+1}^{k+1} (v_i, v_j)$$

$$|A(L_A)| = \frac{(k+2)(k+1)}{2}$$

Note that all the arcs of $A(L_A)$ may be part of the solution $L_A[V']$ of certain present subgraphs $G[V']$; for the arcs of $A \setminus A(L_A)$, no solution for any present subinstance includes any of them. Consequently:

$$d(L'_A) = \sum_{(v_i,v_j)\in A(L_A)} d(v_i,v_j)\, 1_{\{(v_i,v_j)\in L'_A\}} \qquad [4.7]$$

Using [4.7] and the fact that $A(L_A) = L_A \cup (A(L_A) \setminus L_A)$, we get:

$$\begin{aligned}
E\left(G, \vec{d}, L_A, \mathtt{MA}\right) &= \sum_{V'\subseteq V} \Pr[V']\, d(L'_A) \\
&= \sum_{V'\subseteq V} \Pr[V'] \sum_{(v_i,v_j)\in A(L_A)} d(v_i,v_j)\, 1_{\{(v_i,v_j)\in L'_A\}} \\
&= \sum_{V'\subseteq V} \Pr[V'] \left(\sum_{(v_i,v_j)\in L_A} d(v_i,v_j)\, 1_{\{(v_i,v_j)\in L'_A\}} \right. \\
&\quad \left. + \sum_{(v_i,v_j)\in (A(L_A)\setminus L_A)} d(v_i,v_j)\, 1_{\{(v_i,v_j)\in L'_A\}} \right) \\
&= \sum_{(v_i,v_j)\in L_A} \sum_{V'\subseteq V} \Pr[V']\, 1_{\{(v_i,v_j)\in L'_A\}} d(v_i,v_j) \\
&\quad + \left(\sum_{(v_i,v_j)\in (A(L_A)\setminus L_A)} \sum_{V'\subseteq V} \Pr[V']\, 1_{\{(v_i,v_j)\in L'_A\}} d(v_i,v_j) \right) \qquad [4.8]
\end{aligned}$$

In the case where $(v_i, v_j) \in L_A$ (i.e., $(v_i, v_j) = a_i$), $(v_i, v_j) \in L'_A$ only if both v_i and v_j are present in V'; on the other hand, if $(v_i, v_j) \in (A(L_A) \setminus L_A)$, then $(v_i, v_j) \in L'_A$, only if both v_i and v_j are present in V' and, moreover, all vertices v_l, $i < l < j$ are absent. Finally:

$$A(L_A) \setminus L_A = \{(v_i, v_j) : i = 0, \ldots, k-1, j = i+2, \ldots, k+1\} \qquad [4.9]$$

$$\begin{aligned}
|A(L_A) \setminus L_A| &= \sum_{i=0}^{k-1} \sum_{j=i+2}^{k+1} 1 = \sum_{i=0}^{k-1} (k+1-(i+2)+1) \\
&= \sum_{i=0}^{k-1} (k-i) = (k+1)^2 - \sum_{i=0}^{k-1} i = \frac{(k^2+k)}{2} \qquad [4.10]
\end{aligned}$$

Using [4.9] and [4.10], expression [4.8] becomes:

$$\begin{aligned} E\left(G, \vec{d}, L_A, \texttt{MA}\right) &= \sum_{i=0}^{k} p_i p_{i+1} d\left(v_i, v_{i+1}\right) \\ &\quad + \sum_{i=0}^{k-1} \sum_{j=i+2}^{k+1} \sum_{V' \subseteq V} \Pr\left[V'\right] d\left(v_i, v_j\right) 1_{\left\{(v_i, v_j) \in L'_A\right\}} \\ &= \sum_{i=0}^{k} p_i p_{i+1} d\left(v_i, v_{i+1}\right) \\ &\quad + \sum_{i=0}^{k-1} \sum_{j=i+2}^{k+1} p_i p_j \left(\prod_{l=i+1}^{j-1} \left(1 - p_l\right)\right) d\left(v_i, v_j\right) \end{aligned} \quad [4.11]$$

In order to compute righthand side of [4.11], given that $k \leqslant n$, we have to perform, at most, n multiplications for each term of the double sums; since each sum performs, at most, n additions, we conclude that the total computation complexity of $E(G, \vec{d}, L_A, \texttt{MA})$ is of $O(n^3)$.

4.4.3. *An algebraic proof for Theorem 4.3*

From the expression for $E(G, \vec{d}, L_A, \texttt{MA})$ of Theorem 4.2, setting $d(v_i, v_j) = 1$, $v_i, v_j \in V \cup \{s, t\}$, we get:

$$E\left(G, \vec{d}, L_A, \texttt{MA}\right) = \sum_{i=0}^{k} p_i p_{i+1} + \sum_{i=0}^{k-1} \sum_{j=i+2}^{k+1} p_i p_j \left(\prod_{l=i+1}^{j-1} \left(1 - p_l\right)\right) \quad [4.12]$$

Dealing with the righthand side of [4.12], we prove now that:

$$\sum_{i=0}^{k} p_i p_{i+1} + \sum_{i=0}^{k-1} \sum_{j=i+2}^{k+1} p_i p_j \left(\prod_{l=i+1}^{j-1} \left(1 - p_l\right)\right) = 1 + \sum_{i=1}^{k} p_i \quad [4.13]$$

Note first that:

$$\begin{aligned} \sum_{i=0}^{k-1} \sum_{j=i+2}^{k+1} p_i p_j \prod_{l=i+1}^{j-1} \left(1 - p_l\right) &= \sum_{i=0}^{k-1} p_i \sum_{j=i+2}^{k+1} p_j \prod_{l=i+1}^{j-1} \left(1 - p_l\right) \\ &= \sum_{i=0}^{k-1} p_i \left(\sum_{j=i+1}^{k+1} p_j \prod_{l=i+1}^{j-1} \left(1 - p_l\right) - p_{i+1}\right) \end{aligned}$$

Given $i_0 \in \{1, \ldots, k\}$ and k_0 such that $k + 1 \geqslant k_0 \geqslant i_0$, we show by induction on k_0 that:

$$\sum_{j=i_0}^{k_0} p_j \prod_{l=i_0}^{j-1} (1 - p_l) = 1 - \prod_{l=i_0}^{k_0} (1 - p_l) \qquad [4.14]$$

For $k_0 = i_0$, we have $p_{i_0} = 1-(1-p_{i_0})$, so the equality holds. Assume that [4.14] holds for $h \leqslant k$; we shall prove it for $h + 1$:

$$\begin{aligned}
\sum_{j=i_0}^{h+1} p_j \prod_{l=i_0}^{j-1} (1 - p_l) &= \sum_{j=i_0}^{h} p_j \prod_{l=i_0}^{j-1} (1 - p_l) + p_{h+1} \prod_{l=i_0}^{h} (1 - p_l) \\
&= 1 - \prod_{l=i_0}^{h} (1 - p_l) + p_{h+1} \prod_{l=i_0}^{h} (1 - p_l) \\
&= 1 - \prod_{l=i_0}^{h} (1 - p_l) (1 - p_{h+1}) \\
&= 1 - \prod_{l=i_0}^{h+1} (1 - p_l)
\end{aligned}$$

In particular, for $k_0 = k + 1$, $i_0 = i + 1$, $i \in \{0, \ldots, k - 1\}$, we get:

$$\sum_{j=i+1}^{k+1} p_j \prod_{l=i+1}^{j-1} (1 - p_l) = 1 - \prod_{l=i+1}^{k+1} (1 - p_l)$$

Since $p_{k+1} = \Pr[t] = 1$, we have:

$$\sum_{j=i+1}^{k+1} p_j \prod_{l=i+1}^{j-1} (1 - p_l) = 1$$

so:

$$\sum_{i=0}^{k} p_i p_{i+1} + \sum_{i=0}^{k-1} \sum_{j=i+2}^{k+1} p_i p_j \prod_{l=i+1}^{j-1} (1 - p_l) \qquad [4.15]$$

$$
\begin{aligned}
&= \sum_{i=0}^{k} p_i p_{i+1} + \sum_{i=0}^{k-1} p_i \left(1 - p_{i+1}\right) \\
&= p_k p_{k+1} + \sum_{i=0}^{k-1} p_i p_{i+1} + \sum_{i=0}^{k-1} p_i - \sum_{i=0}^{k-1} p_i p_{i+1} \\
&= p_k + \sum_{i=0}^{k-1} p_i \;=\; \sum_{i=0}^{k} p_i \\
&= p_0 + \sum_{i=1}^{k} p_i \;=\; 1 + \sum_{i=1}^{k} p_i
\end{aligned}
\qquad [4.16]
$$

and this concludes [4.13] and the algebraic proof of the expression for $E(G, L_A, \texttt{MA})$ of Theorem 4.3.

As we have already noted at the end of section 4.2.1, [4.15] establishes the following corollary.

COROLLARY 4.3.– Consider $n+2$ numbers $x_i \leqslant 1$, $i = 0, \ldots, n+1$ with $x_0 = x_{n+1} = 1$. Then:

$$
\sum_{i=0}^{n} x_i x_{i+1} + \sum_{i=0}^{n-1} \sum_{j=i+2}^{n+1} x_i x_j \left(\prod_{l=i+1}^{j-1} \left(1 - x_l\right) \right) = 1 + \sum_{i=1}^{n} x_i
$$

4.4.4. *Proof of Lemma 4.1*

The proof is done, for fixed i, by induction on m:

– for $m = i$: $x_i = 1 - (1 - x_i)$ and the expression claimed is true;

– assume that the expression is true at range m;

– then, at range $m+1$, we have:

$$
\begin{aligned}
\sum_{j=i}^{m+1} x_j \prod_{l=i}^{j-1} \left(1 - x_l\right) &= \sum_{j=i}^{m} x_j \prod_{l=i}^{j-1} \left(1 - x_l\right) + x_{m+1} \prod_{l=i}^{m} \left(1 - x_l\right) \\
&= 1 - \prod_{l=i}^{m} \left(1 - x_l\right) + x_{m+1} \prod_{l=i}^{m} \left(1 - x_l\right) \\
&= 1 - \prod_{l=i}^{m} \left(1 - x_l\right)\left(1 - x_{m+1}\right) = 1 - \prod_{l=i}^{m+1} \left(1 - x_l\right)
\end{aligned}
$$

This completes the proof of the lemma.

4.4.5. *Proof of Lemma 4.2*

Given that $x_n \leqslant 1$, in order to prove the claimed inequality, it suffices to prove that:

$$\sum_{i=0}^{n-2} x_i \prod_{j=i+1}^{n-1} (1 - x_j) \leqslant 1 - x_{n-1}$$

Let us consider the series $U_l = \sum_{i=0}^{l-1} x_i \prod_{j=i+1}^{l} (1 - x_j)$, $1 \leqslant l \leqslant n$.

We shall prove that $U_{n-1} \leqslant 1 - x_{n-1}$.

Series U_l is recurrently defined as $U_{l+1} = (1 - x_{l+1})(U_l + x_l)$. In fact:

$$\begin{aligned} U_{l+1} &= \sum_{i=0}^{l} x_i \prod_{j=i+1}^{l+1} (1 - x_j) = \sum_{i=0}^{l-1} x_i \prod_{j=i+1}^{l+1} (1 - x_j) + x_l (1 - x_{l+1}) \\ &= \sum_{i=0}^{l-1} x_i (1 - x_{l+1}) \prod_{j=i+1}^{l} (1 - x_j) + x_l (1 - x_{l+1}) \\ &= (1 - x_{l+1})(U_l + x_l) \end{aligned}$$

We now prove by induction on l that $U_l \leqslant 1 - x_l$:

– for $l = 1$: $U_1 = x_0(1 - x_1) \leqslant 1 - x_1$ (because $x_0 \leqslant 1$);

– suppose that the inequality claimed is true at range l;

– at range $l + 1$, we get:

$$U_{l+1} = (1 - x_{l+1})(U_l + x_l) \leqslant (1 - x_{l+1})(1 - x_l + x_l) \leqslant 1 - x_{l+1}$$

and the proof of the lemma is complete.

4.4.6. *Proof of Theorem 4.4*

Starting from the generic expression for $E(G, \vec{d}, L_A, \mathtt{MA}) = E(G, \vec{d}, \mu, \mathtt{MA})$ given in Theorem 4.2, we will re-write it in order to fit $E(G, \vec{d}, \mu', \mathtt{MA})$:

$$E\left(G, \vec{d}, \mu', \mathtt{MA}\right) = \sum_{i=0}^{i_0-1} p_i p_{i+1} d\left(v_i, v_{i+1}\right) + p_{i_0} p_{i_2} d\left(v_{i_0}, v_{i_2}\right)$$

$$
\begin{aligned}
&+\sum_{i=i_2}^{k} p_i p_{i+1} d(v_i, v_{i+1}) + \sum_{\substack{i=0\\ i\neq i_1}}^{k-1} \sum_{\substack{j=i+2\\ j\neq i_1}}^{k+1} p_i p_j \prod_{\substack{l=i+1\\ l\neq i_1}}^{j-1} (1-p_l)\, d(v_i, v_j) \\
=\;& \sum_{i=0}^{i_0-1} p_i p_{i+1} d(v_i, v_{i+1}) + p_{i_0} p_{i_2} d(v_{i_0}, v_{i_2}) \\
&+\sum_{i=i_2}^{k} p_i p_{i+1} d(v_i, v_{i+1}) + \sum_{i=0}^{i_1-1=i_0} \sum_{\substack{j=i+2\\ j\neq i_1}}^{k+1} p_i p_j \prod_{\substack{l=i+1\\ l\neq i_1}}^{j-1} (1-p_l)\, d(v_i, v_j) \\
&+ \sum_{i=i_1+1=i_2}^{k-1} \sum_{j=i+2}^{k+1} p_i p_j \prod_{l=i+1}^{j-1} (1-p_l)\, d(v_i, v_j) \\
=\;& \sum_{i=0}^{i_0-1} p_i p_{i+1} d(v_i, v_{i+1}) + p_{i_0} p_{i_2} d(v_{i_0}, v_{i_2}) + \sum_{i=i_2}^{k} p_i p_{i+1} d(v_i, v_{i+1}) \\
&+\sum_{i=0}^{i_0-2} \sum_{\substack{j=i+2\\ j\neq i_1}}^{k+1} p_i p_j \prod_{\substack{l=i+1\\ l\neq i_1}}^{j-1} (1-p_l)\, d(v_i, v_j) \\
&+ p_{i_0-1} \sum_{j=i_2}^{k+1} p_j \prod_{\substack{l=i_0\\ l\neq i_1}}^{j-1} (1-p_l)\, d(v_{i_0-1}, v_j) \\
&+ p_{i_0} \sum_{j=i_2+1}^{k+1} p_j \prod_{l=i_2}^{j-1} (1-p_l)\, d(v_{i_0}, v_j) \\
&+\sum_{i=i_2}^{k-1} \sum_{j=i+2}^{k+1} p_i p_j \prod_{l=i+1}^{j-1} (1-p_l)\, d(v_i, v_j) \\
=\;& \sum_{i=0}^{i_0-1} p_i p_{i+1} d(v_i, v_{i+1}) + p_{i_0} p_{i_2} d(v_{i_0}, v_{i_2}) + \sum_{i=i_2}^{k} p_i p_{i+1} d(v_i, v_{i+1}) \\
&+\sum_{i=0}^{i_0-2} \sum_{j=i+2}^{i_0} p_i p_j \prod_{l=i+1}^{j-1} (1-p_l)\, d(v_i, v_j) \\
&+\sum_{i=0}^{i_0-2} \sum_{j=i_2}^{k+1} p_i p_j \prod_{\substack{l=i+1\\ l\neq i_1}}^{j-1} (1-p_l)\, d(v_i, v_j)
\end{aligned}
$$

$$+ p_{i_0-1} \sum_{j=i_2}^{k+1} p_j \prod_{\substack{l=i_0 \\ l \neq i_1}}^{j-1} (1 - p_l)\, d(v_{i_0-1}, v_j)$$

$$+ p_{i_0} \sum_{j=i_2+1}^{k+1} p_j \prod_{l=i_2}^{j-1} (1 - p_l)\, d(v_{i_0}, v_j)$$

$$+ \sum_{i=i_2}^{k-1} \sum_{j=i+2}^{k+1} p_i p_j \prod_{l=i+1}^{j-1} (1 - p_l)\, d(v_i, v_j) \qquad [4.17]$$

We shall now rewrite the expression for $E(G, \vec{d}, \mu, \mathtt{MA})$ in order to isolate the terms relative to v_{i_1}:

$$E\left(G, \vec{d}, \mu, \mathtt{MA}\right) = \sum_{i=0}^{k} p_i p_{i+1} d(v_i, v_{i+1})$$

$$+ \sum_{i=0}^{k-1} \sum_{j=i+2}^{k+1} p_i p_j \prod_{l=i+1}^{j-1} (1 - p_l)\, d(v_i, v_j)$$

$$= \sum_{i=0}^{i_0-1} p_i p_{i+1} d(v_i, v_{i+1}) + p_{i_0} p_{i_1} d(v_{i_0}, v_{i_1}) + p_{i_1} p_{i_2} d(v_{i_1}, v_{i_2})$$

$$+ \sum_{i=i_2}^{k} p_i p_{i+1} d(v_i, v_{i+1}) + \sum_{i=0}^{i_0-2} \sum_{j=i+2}^{i_0} p_i p_j \prod_{l=i+1}^{j-1} (1 - p_l)\, d(v_i, v_j)$$

$$+ \sum_{i=0}^{i_0-2} \sum_{j=i_1}^{k+1} p_i p_j \prod_{l=i+1}^{j-1} (1 - p_l)\, d(v_i, v_j)$$

$$+ p_{i_0-1} \sum_{j=i_1}^{k+1} p_j \prod_{l=i_0}^{j-1} (1 - p_l)\, d(v_{i_0-1}, v_j)$$

$$+ p_{i_0} \sum_{j=i_2}^{k+1} p_j \prod_{l=i_1}^{j-1} (1 - p_l)\, d(v_{i_0-1}, v_j)$$

$$+ \sum_{i=i_1}^{k-1} \sum_{j=i+2}^{k+1} p_i p_j \prod_{l=i+1}^{j-1} (1 - p_l)\, d(v_i, v_j)$$

$$= \sum_{i=0}^{i_0-1} p_i p_{i+1} d(v_i, v_{i+1}) + p_{i_0} p_{i_1} d(v_{i_0}, v_{i_1}) + p_{i_1} p_{i_2} d(v_{i_1}, v_{i_2})$$

$$+\sum_{i=i_2}^{k} p_i p_{i+1} d(v_i, v_{i+1}) + \sum_{i=0}^{i_0-2} \sum_{j=i+2}^{i_0} p_i p_j \prod_{l=i+1}^{j-1} (1-p_l) d(v_i, v_j)$$

$$+\sum_{i=0}^{i_0-2} p_i p_{i_1} \prod_{l=i+1}^{i_0} (1-p_l) d(v_i, v_{i_1})$$

$$+\sum_{i=0}^{i_0-2} \sum_{j=i_2}^{k+1} p_i p_j \prod_{l=i+1}^{j-1} (1-p_l) d(v_i, v_j)$$

$$+p_{i_0-1}\left(p_{i_1}(1-p_{i_0}) d(v_{i_0-1}, v_{i_1})\right.$$

$$\left.+\sum_{j=i_2}^{k+1} p_j \prod_{l=i_0}^{j-1} (1-p_l) d(v_{i_0-1}, v_j)\right)$$

$$+p_{i_0}\left(p_{i_2}(1-p_{i_1}) d(v_{i_0}, v_{i_2})\right.$$

$$\left.+\sum_{j=i_2+1}^{k+1} p_j \prod_{l=i_1}^{j-1} (1-p_l) d(v_{i_0}, v_j)\right)$$

$$+p_{i_1} \sum_{j=i_1+2}^{k+1} p_j \prod_{l=i_2}^{j-1} (1-p_l) d(v_{i_1}, v_j)$$

$$+\sum_{i=i_2}^{k-1} \sum_{j=i+2}^{k+1} p_i p_j \prod_{l=i+1}^{j-1} (1-p_l) d(v_i, v_j) \qquad [4.18]$$

From [4.17] and [4.18] for $E(G, \vec{d}, \mu, \mathtt{MA})$ and $E(G, \vec{d}, \mu', \mathtt{MA})$, respectively, we get for their difference, denoted by $D(\mu, \mu')$:

$$D(\mu, \mu') = E\left(G, \vec{d}, \mu, \mathtt{MA}\right) - E\left(G, \vec{d}, \mu', \mathtt{MA}\right)$$

$$= p_{i_0} p_{i_1} d(v_{i_0}, v_{i_1}) + p_{i_1} p_{i_2} d(v_{i_1}, v_{i_2})$$

$$- p_{i_0} p_{i_2} d(v_{i_0}, v_{i_2}) + p_{i_1} \sum_{i=0}^{i_0-2} p_i \prod_{l=i+1}^{i_0} (1-p_l) d(v_i, v_{i_1})$$

$$+\sum_{i=0}^{i_0-2} \sum_{j=i_2}^{k+1} p_i p_j \left(\prod_{l=i+1}^{j-1} (1-p_l) - \prod_{\substack{l=i+1 \\ l \neq i_1}}^{j-1} (1-p_l)\right) d(v_i, v_j)$$

$$+ p_{i_0-1}p_{i_1}(1-p_{i_0})\,d(v_{i_0-1},v_{i_1})$$

$$+ p_{i_0-1}\sum_{j=i_2}^{k+1} p_j \left(\prod_{l=i_0}^{j-1}(1-p_l) - \prod_{\substack{l=i_0\\ l\neq i_1}}^{j-1}(1-p_l)\right) d(v_{i_0-1},v_j)$$

$$+ p_{i_0}p_{i_2}(1-p_{i_1})\,d(v_{i_0},v_{i_2})$$

$$+ p_{i_0}\sum_{j=i_2+1}^{k+1} p_j \left(\prod_{l=i_1}^{j-1}(1-p_l) - \prod_{l=i_2}^{j-1}(1-p_l)\right) d(v_{i_0},v_j)$$

$$+ p_{i_1}\sum_{j=i_1+2}^{k+1} p_j \prod_{l=i_2}^{j-1}(1-p_l)\,d(v_{i-1},v_j) \qquad [4.19]$$

Since

$$\prod_{l=i+1}^{j-1}(1-p_l) - \prod_{l=i+1,l\neq i_1}^{j-1}(1-p_l) = -p_{i_1}\prod_{l=i+1,l\neq i_1}^{j-1}(1-p_l)$$

quantity $D(\mu,\mu')$ (expression [4.19]) becomes:

$$D(\mu,\mu') = p_{i_0}p_{i_1}d(v_{i_0},v_{i_1}) + p_{i_1}p_{i_2}d(v_{i_1},v_{i_2})$$

$$- p_{i_0}p_{i_2}d(v_{i_0},v_{i_2}) + p_{i_1}\sum_{i=0}^{i_0-2} p_i \prod_{l=i+1}^{i_0}(1-p_l)\,d(v_i,v_{i_1})$$

$$- p_{i_1}\sum_{i=0}^{i_0-2}\sum_{j=i_2}^{k+1} p_i p_j \prod_{\substack{l=i+1\\ l\neq i_1}}^{j-1}(1-p_l)\,d(v_i,v_j)$$

$$+ p_{i_0-1}p_{i_1}(1-p_{i_0})\,d(v_{i_0-1},v_{i_1})$$

$$- p_{i_0-1}p_{i_1}\sum_{j=i_2}^{k+1} p_j \prod_{\substack{l=i_0\\ l\neq i_1}}^{j-1}(1-p_l)\,d(v_{i_0-1},v_j)$$

$$+ p_{i_0}p_{i_2}d(v_{i_0},v_{i_2}) - p_{i_0}p_{i_2}p_{i_1}d(v_{i_0},v_{i_2})$$

$$- p_{i_0}p_{i_1}\sum_{j=i_2+1}^{k+1} p_j \prod_{l=i_2}^{j-1}(1-p_l)\,d(v_{i_0},v_j)$$

$$+ p_{i_1}\sum_{j=i_1+2}^{k+1} p_j \prod_{l=i_2}^{j-1}(1-p_l)\,d(v_{i_1},v_j) \qquad [4.20]$$

From [4.20], since the terms not containing p_{i_1} are mutually cancelled, we get:

$$\begin{aligned}
\frac{D(\mu,\mu')}{p_{i_1}} &= p_{i_0} d(v_{i_0}, v_{i_1}) + p_{i_2} d(v_{i_1}, v_{i_2}) \\
&+ \sum_{i=0}^{i_0-2} p_i \prod_{l=i+1}^{i_0} (1-p_l)\, d(v_i, v_{i_1}) \\
&- \sum_{i=0}^{i_0-2} \sum_{j=i_2}^{k+1} p_i p_j \prod_{\substack{l=i+1 \\ l\neq i_1}}^{j-1} (1-p_l)\, d(v_i, v_j) \\
&+ p_{i_0-1}(1-p_{i_0})\, d(v_{i_0-1}, v_{i_1}) - p_{i_0-1} p_{i_2} (1-p_{i_0})\, d(v_{i_0-1}, v_{i_2}) \\
&- p_{i_0-1} \sum_{j=i_2+1}^{k+1} p_j \prod_{\substack{l=i_0 \\ l\neq i_1}}^{j-1} (1-p_l)\, d(v_{i_0-1}, v_j) \\
&- p_{i_0} p_{i_2} d(v_{i_0}, v_{i_2}) - p_{i_0} \sum_{j=i_2+1}^{k+1} p_j \prod_{l=i_2}^{j-1} (1-p_l)\, d(v_{i_0}, v_j) \\
&+ \sum_{j=i_1+2}^{k+1} p_j \prod_{l=i_2}^{j-1} (1-p_l)\, d(v_{i_1}, v_j)
\end{aligned}$$

Since $p_{i_1} \geqslant 0$, we must show that $D(\mu,\mu')/p_{i_1} \geqslant 0$. For this, using the fact that G is metric, we will first prove a lower bound for $D(\mu,\mu')/p_{i_1}$. In particular, let us note that for all $d(v_i, v_j)$ with $i < i_1$ and $j > i_1$, we have: $-d(v_i, v_j) \geqslant -d(v_i, v_{i_1}) - d(v_{i_1}, v_j)$. By applying this form of triangular inequality, we get:

$$\begin{aligned}
\frac{D(\mu,\mu')}{p_{i_1}} &\geqslant p_{i_0} d(v_{i_0}, v_{i_1}) + p_{i_2} d(v_{i_1}, v_{i_2}) \\
&+ \sum_{i=0}^{i_0-2} p_i \prod_{l=i+1}^{i_0} (1-p_l)\, d(v_i, v_{i_1}) \\
&- \sum_{i=0}^{i_0-2} \sum_{j=i_2}^{k+1} p_i p_j \prod_{\substack{l=i+1 \\ l\neq i_1}}^{j-1} (1-p_l) \left(d(v_i, v_{i_1}) + d(v_{i_1}, v_j)\right) \\
&+ p_{i_0-1}(1-p_{i_0})\, d(v_{i_0-1}, v_{i_1}) - p_{i_0-1} p_{i_2} (1-p_{i_0})\, d(v_{i_0-1}, v_{i_1}) \\
&- p_{i_0-1} p_{i_2} (1-p_{i_0})\, d(v_{i_1}, v_{i_2})
\end{aligned}$$

$$
\begin{aligned}
&- p_{i_0-1} \sum_{j=i_2+1}^{k+1} p_j \prod_{\substack{l=i_0 \\ l \neq i_1}}^{j-1} (1-p_l) \left(d\left(v_{i_0-1}, v_{i_1}\right) + d\left(v_{i_1}, v_j\right) \right) \\
&- p_{i_0} p_{i_2} d\left(v_{i_0}, v_{i_1}\right) - p_{i_0} p_{i_2} d\left(v_{i_1}, v_{i_2}\right) \\
&- p_{i_0} \sum_{j=i_2+1}^{k+1} p_j \prod_{l=i_2}^{j-1} (1-p_l) \left(d\left(v_{i_0}, v_{i_1}\right) + d\left(v_{i_1}, v_j\right) \right) \\
&+ \sum_{j=i_2+1}^{k+1} p_j \prod_{l=i_2}^{j-1} (1-p_l)\, d\left(v_{i_1}, v_j\right) \\
= \quad & \sum_{i=0}^{i_0-2} p_i d\left(v_i, v_{i_1}\right) \left(\prod_{l=i+1}^{i_0} (1-p_l) - \sum_{j=i_2}^{k+1} p_j \prod_{\substack{l=i+1 \\ l \neq i_1}}^{j-1} (1-p_l) \right) \\
&+ p_{i_0-1} d\left(v_{i_0-1}, v_{i_1}\right) \\
&\times \left(\left(1-p_{i_0}\right) - \left(1-p_{i_0}\right) p_{i_2} - \sum_{j=i_2+1}^{k+1} p_j \prod_{\substack{l=i_0 \\ l \neq i_1}}^{j-1} (1-p_l) \right) \\
&+ p_{i_0} d\left(v_{i_0}, v_{i_1}\right) \left(1 - p_{i_2} - \sum_{j=i_2+1}^{k+1} p_j \prod_{l=i_2}^{j-1} (1-p_l) \right) \\
&+ p_{i_2} d\left(v_{i_1}, v_{i_2}\right) \\
&\times \left(1 - p_{i_0-1}\left(1-p_{i_0}\right) - p_{i_0} - \sum_{i=0}^{i_0-2} p_i \prod_{\substack{l=i+1 \\ l \neq i_1}}^{i_1} (1-p_l) \right) \\
&+ \sum_{j=i_2+1}^{k+1} p_j d\left(v_{i_1}, v_j\right) \\
&\times \left(\left(1-p_{i_0}\right) \prod_{l=i_2}^{j-1} (1-p_l) - p_{i_0-1} \prod_{\substack{l=i_0 \\ l \neq i_1}}^{j-1} (1-p_l) \right. \\
&\left. - \sum_{i=0}^{i_0-2} p_i \prod_{\substack{l=i+1 \\ l \neq i_1}}^{j-1} (1-p_l) \right)
\end{aligned}
$$

$$
\begin{aligned}
= & \sum_{i=0}^{i_0-2} p_i d\left(v_i, v_{i_1}\right) \\
& \times \left(\prod_{l=i+1}^{i_0} (1-p_l)\right) \left(1 - \sum_{j=i_2}^{k+1} p_j \prod_{l=i_2}^{j-1} (1-p_l)\right) \\
& + p_{i_0-1} d\left(v_{i_0-1}, v_{i_1}\right)\left(1-p_{i_0}\right) \\
& \times \left(1 - p_{i_2} - \sum_{j=i_2+1}^{k+1} p_j \prod_{l=i_2}^{j-1} (1-p_l)\right) \\
& + p_{i_0} d\left(v_{i_0}, v_{i_1}\right) \left(1 - p_{i_2} - \sum_{j=i_2+1}^{k+1} p_j \prod_{l=i_2}^{j-1} (1-p_l)\right) \\
& + p_{i_2} d\left(v_{i_1}, v_{i_2}\right) \\
& \times \left(\left(1-p_{i_0-1}\right)\left(1-p_{i_0}\right) - \sum_{i=0}^{i_0-2} p_i \prod_{l=i+1}^{i_0} (1-p_l)\right) \\
& + \sum_{j=i_2+1}^{k+1} p_j d\left(v_{i_1}, v_j\right) \left(\prod_{l=i_2}^{j-1} (1-p_l)\right) \\
& \times \left(\left(1-p_{i_0}\right)\left(1-p_{i_0-1}\right) - \sum_{i=0}^{i_0-2} p_i \prod_{l=i+1}^{i_0} (1-p_l)\right)
\end{aligned} \qquad [4.21]
$$

To complete the proof of the theorem, it suffices to show that the lower bound for $D(\mu, \mu')/p_{i_1}$ provided in [4.21] is non-negative. Indeed:

– for the factor of $d(v_i, v_{i_1})$, by Lemma 4.1 and taking into account that $p_{k+1} = 1$, we get: $1 - \sum_{j=i_2}^{k+1} p_j \prod_{l=i_2}^{j-1}(1-p_l) = \prod_{l=i_2}^{k+1}(1-p_l) = 0$;

– for the terms multiplying $d(v_{i_0-1}, v_{i_1})$ and $d(v_{i_0}, v_{i_1})$, respectively, we have by Lemma 4.1:

$$
1 - p_{i_2} - \sum_{j=i_2+1}^{k+1} p_j \prod_{l=i_2}^{j-1} (1-p_l) =
$$

$$
1 - p_{i_2} - \left(\sum_{j=i_2}^{k+1} p_j \prod_{l=i_2}^{j-1} (1-p_l) - p_{i_2}\right) = \prod_{l=i_2}^{k+1} (1-p_l) = 0
$$

– finally, by direct application of Lemma 4.2, the two last terms are positive.

Chapter 5

Probabilistic Minimum Coloring

We study in this chapter PROBABILISTIC MIN COLORING under the simple modification strategy M consisting, *given an a priori solution C, of removing the absent vertices from C* (this is strategy MS of Chapter 1, fitting PROBABILISTIC MIN COLORING). As we will see in section 5.1, PROBABILISTIC MIN COLORING becomes in our probabilistic framework a kind of weighted coloring and the objective then is to determine such a coloring of minimum total-weight such coloring. We deal with two versions of this problem:

– the first one is the natural version where one wishes to determine a best (minimum weight) coloring;

– the second version, called PROBABILISTIC MIN k-COLORING, consists, given $k \in \mathbb{N}$ of determining a minimum-value k-coloring of the graph, i.e., a best coloring using exactly k colors.

PSfrag replacements

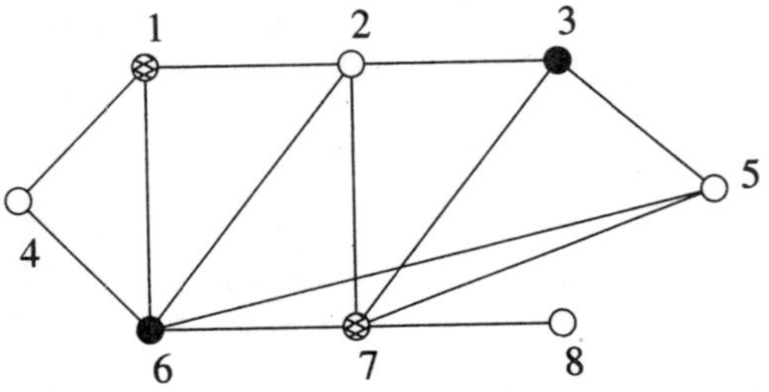

Figure 5.1. *A graph G, together with the a priori coloring (white, black, shadow)*

For both versions, we settle complexity and approximation issues in general graphs as well as in natural particular classes of graphs as bipartite graphs, split graphs, etc.

EXAMPLE 5.1.– Consider the graph G of Figure 5.1, together with the *a priori* 3-coloring $C = (\{2,4,5,8\},\{3,6\},\{1,7\})$ (white, black, shadow) and assume that vertices 1, 7 and 8 are absent. Application of MS in G' produces the (white, black) 2-coloring $C' = (\{2,4,5\},\{3,6\})$ (Figure 5.2).

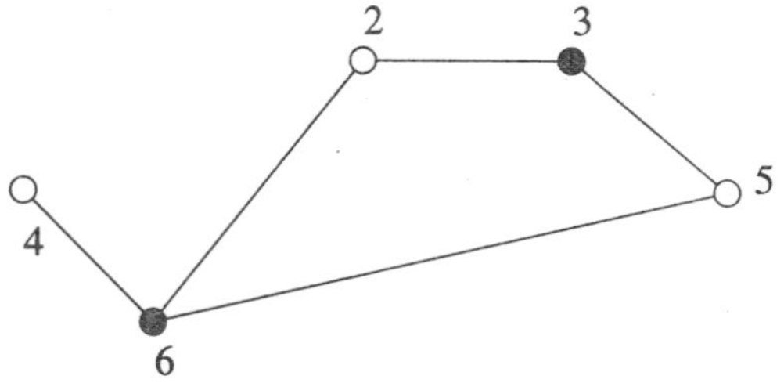

Figure 5.2. *Application of strategy MS on G' (Figure 5.1), starting from the a priori (white, black, shadow) coloring; colors produced: (white, black)*

Given an *a priori* coloring $C = (S_1, \ldots, S_k)$ for V, the functional to be minimized is defined as: $E(G, C, \mathtt{M}) = \sum_{V' \subseteq V} \Pr[V']|C(V', \mathtt{M})|$, where $C(V', \mathtt{M})$ is the solution computed by $\mathtt{M}(C, V')$ (i.e., by M when executed with inputs the *a priori* solution C and $G[V']$) and, as usually, $\Pr[V'] = \prod_{i \in V'} p_i \prod_{i \in V \setminus V'} (1 - p_i)$.

In what follows, given a graph $G(V, E)$, we denote by C^* an optimal *a priori* coloring of G. Furthermore, since the modification strategy M is fixed for the rest of the chapter, we will simplify notations by using $E(G, C)$ and $E(G, C^*)$ instead of $E(G, C, \mathtt{M})$ and $E(G, C^*, \mathtt{M})$, respectively, and $C(V')$ instead of $C(V', \mathtt{M})$. Finally, we shall denote by $p_{\max}$ (resp., $p_{\min}$) the maximum (resp., minimum) vertex-probability of V.

We have already seen in the Example 1.2 of Chapter 1 how a natural timetabling problem can be modeled and faced as a probabilistic coloring problem. As another example, consider a planning aiding process for realizing satellite shots. We associate a vertex with any shot requested and we link two vertices if they correspond to shots that cannot be realized by the satellite on the same orbit. But a shot realized under, for example, strong cloud covering cannot be used for the purposes for which it has been requested. Using meteorological forecasting, one can assign to any shot requested a probability that it will be usable. For a mean-term planning, one of the main problems is to decide if, on a given time slot (consequently, for a fixed number of orbits), a shot can or cannot be realized. It has been shown ([GAB 97]) that this problem can be modeled as a minimum partition into cliques of the vertices of the graph outlined above. So, if one takes into account probabilities associated with meteorological forecasting, one has to solve a probabilistic version of the problem mentioned. Note that the partition into cliques in a graph amounts to a coloring problem in its complement.

5.1. The functional $E(G,C)$

In this section, we first analytically express the functional for PROBABILISTIC MIN COLORING; then, based on it, we show that it can be computed in polynomial time. Moreover, always based upon the analytical expression obtained for the functional, we give a combinatorial characterization of the optimal *a priori* solution.

Recall that given an *a priori* solution $C = (S_1, S_2, \dots, S_k)$ of cardinality k and a subgraph $G' = G[V']$ of G, we denote by $C(V')$ the coloring of G' obtained by restriction of C in V' and set $k' = |C(V')|$. As has already been noted, $E(G,C) = \sum_{V' \subseteq V} \mathbf{Pr}[V']|C(V')|$. Using the notations just above:

$$E(G,C) = \sum_{V' \subseteq V} \mathbf{Pr}\left[V'\right]\left|C\left(V'\right)\right| = \sum_{V' \subseteq V} \mathbf{Pr}\left[V'\right] k' \qquad [5.1]$$

Consider variable x_j defined by:

$$x_j = \begin{cases} 1 & \text{if } S_j \cap V' = \emptyset \\ 0 & \text{otherwise} \end{cases}$$

Note that condition $S_j \cap V' = \emptyset$ means that color S_j is empty in G' (i.e., that the vertices of S_j are absent from V'). Then k' can be written as $k' = \sum_{j=1}^{k}(1 - x_j)$ and [5.1] becomes:

$$\begin{aligned} E(G,C) &= \sum_{V' \subseteq V} \mathbf{Pr}\left[V'\right]\left(\sum_{j=1}^{k}(1 - x_j)\right) \\ &= \sum_{V' \subseteq V} \mathbf{Pr}\left[V'\right]\sum_{j=1}^{k} 1 - \sum_{V' \subseteq V} \mathbf{Pr}\left[V'\right]\sum_{j=1}^{k} x_j \\ &= \sum_{j=1}^{k}\sum_{V' \subseteq V} \mathbf{Pr}\left[V'\right] - \sum_{j=1}^{k}\sum_{V' \subseteq V} \mathbf{Pr}\left[V'\right] x_j \\ &= k - \sum_{j=1}^{k}\prod_{v_i \in S_j}(1 - p_i) = \sum_{j=1}^{k}\left(1 - \prod_{v_i \in S_j}(1 - p_i)\right) \qquad [5.2] \end{aligned}$$

It is easy to see that computation of $E(G,C)$ needs at most $O(n^2)$ arithmetical operations.

Notice that [5.2] provides a closed characterization of the optimal *a priori* solution C^* for PROBABILISTIC MIN COLORING: if the value of an independent set S_j of G is $1 - \prod_{v_i \in S_j}(1 - p_i)$, then *the optimal a priori coloring for G is the partition into independent sets for which the sum of their values is the smallest over all such partitions*. So, as in the case of the problems covered in the previous chapters of this book under strategy MS, PROBABILISTIC MIN COLORING can be equivalently stated as a "deterministic combinatorial optimization problem" as follows: *given a graph $G(V,E)$, and a vertex-probability vector* $\mathbf{Pr}$, *determine a coloring $C^* = (S_1^*, S_2^*, \ldots)$ minimizing quantity $f(G, C^*, \mathbf{Pr}) = \sum_{S_j^* \in C^*}(1 - \prod_{v_i \in S_j^*}(1 - p_i))$, where $p_i = \mathbf{Pr}[v_i]$ denotes the probability of vertex $v_i \in V$.*

It can be immediately noted that PROBABILISTIC MIN COLORING is completely different from the problems studied in the previous chapters. There, when absent vertices were dropped out of the *a priori* solutions considered, the optimal *a priori* solutions were a maximum weight independent set, or a minimum weight vertex-covering, of the input graph considering that its vertices receive their probabilities as weights. Here, the problem covered is not so simple, since weight of an independent set is not an additive function. Let us note that there exist several weighted versions of the minimum coloring. For example, one can consider that the weight of a color is the maximum (or the minimum, or even the average) weight of the vertices in the independent set representing it, and the objective is to find a coloring minimizing the sum of the weights of the colors (see, for example, [DEM 01, DEM 02] for a version of weighted coloring, where the weight of a color is the maximum over the weights of its vertices). Also, another problem that could be seen as a non-standard version of weighted coloring is the so-called "chromatic sum" problem considered in [BAR 88, JAN 97, NIC 99]. The problem of this chapter is quite different from all these coloring-versions. Note also that minimum coloring is one of the paradigmatic problems for the combinatorial optimization and one of the hardest ones from both optimal and approximated solutions perspectives (see, for example, [GAR 79] for the former and [AUS 99, PAS 03, VAZ 01] for the latter). Yet for this reason, it is always interesting to apprehend it under several objective functions in order to better capture its intractability facets.

We now provide upper and lower bounds on the value of $E(G,C)$ valid for any graph. They will be used later for achieving approximation results about PROBABILISTIC MIN COLORING.

Consider a graph $G(V, E)$, a coloring $C = (S_1, \ldots, S_k)$, and set $n = |V|$. Then:

$$\begin{aligned} E(G,C) &\leqslant \sum_{i=1}^{n} p_i \\ E(G,C) &\geqslant 1 - \prod_{v_i \in V} (1-p_i) \geqslant \sum_{i=1}^{n} p_i - \sum_{i=1}^{n} \sum_{j=i+1}^{n} p_i p_j \end{aligned} \qquad [5.3]$$

$$\begin{aligned} E(G,C) &\leqslant \min\{k, np_{\max}\} \\ E(G,C) &\geqslant \max\left\{ \sum_{j=1}^{k} \left(1 - \exp\left\{ - \sum_{v_i \in S_j} p_i \right\}\right), kp_{\min} \right\} \end{aligned} \qquad [5.4]$$

We first prove the upper bound of [5.3] and the rightmost lower bound. Consider a coloring $C = (S_1, \ldots, S_k)$. We first produce a framing for the term $1 - \prod_{v_i \in S_j}(1 - p_i)$. For simplicity, assume $|S_j| = \ell$ and arbitrarily denote vertices in S_j by $v_1, \ldots, v_\ell$. Then, by induction in ℓ, the following holds:

$$\sum_{i=1}^{\ell} p_i - \sum_{i=1}^{\ell} \sum_{j=i+1}^{\ell} p_i p_j \leqslant 1 - \prod_{i=1}^{\ell} (1-p_i) \leqslant \sum_{i=1}^{\ell} p_i \qquad [5.5]$$

The proof of [5.5] is given in section 5.9.1.

Taking the sums of the members of [5.5] for $m = 1$ to k, the right-hand side inequality immediately gives $E(G, C) \leqslant \sum_{i=1}^{n} p_i$.

We now prove that $E(G, C) \geqslant \sum_{i=1}^{n} p_i - \sum_{i=1}^{n} \sum_{j=i+1}^{n} p_i p_j$, i.e., the rightmost lower bound claimed in [5.3]. From the left-hand side of [5.5], we get:

$$\begin{aligned} \sum_{m=1}^{k} \left(\sum_{i=1}^{\ell} p_i - \sum_{i=1}^{\ell} \sum_{j=i+1}^{\ell} p_i p_j \right) &= \sum_{i=1}^{n} p_i - \sum_{m=1}^{k} \sum_{i=1}^{\ell} \sum_{j=i+1}^{\ell} p_i p_j \\ &\geqslant \sum_{i=1}^{n} p_i - \sum_{i=1}^{n} \sum_{j=i+1}^{n} p_i p_j \end{aligned} \qquad [5.6]$$

Observe that, from the first inequality of [5.5], we have:

$$\sum_{m=1}^{k} \left(\sum_{i=1}^{\ell} p_i - \sum_{i=1}^{\ell} \sum_{j=i+1}^{\ell} p_i p_j \right) \leqslant \sum_{m=1}^{k} \left(1 - \prod_{i=1}^{\ell} (1-p_i) \right) \qquad [5.7]$$

The righthand side of [5.7] is exactly $E(G, C)$. Putting this together with [5.6], the rightmost lower bound for $E(G, C)$ in [5.3] is proved.

In order to prove the second inequality in [5.3], just apply the first inequality in [5.5] setting $\ell = n$.

We finally prove that $1 - \prod_{v_i \in V}(1 - p_i) \leqslant E(G, C)$. For this, we first recall that given three numbers $p_1, p_2, q \in \mathbb{R}^+$ such that $p_1 \leqslant p_2$ and $q \leqslant 1$, then:

$$1 - p_1 q + 1 - \frac{p_2}{q} \leqslant 1 - p_1 + 1 - p_2 \quad [5.8]$$

Expression [5.8] is equivalent to $p_1 q + p_2/q \geqslant p_1 + p_2$. Starting from this expression, some very simple algebra leads to $(1 - q)(p_1 - (p_2/q)) \leqslant 0$, which is true since $1 - q \geqslant 0$ and $p_1 \leqslant p_2 \leqslant p_2/q$.

Consider two colors S_i and S_j of C, set $g(S_i) = \prod_{v_k \in S_i}(1 - p_k)$ and $g(S_j) = \prod_{v_k \in S_j}(1 - p_k)$ and assume that $g(S_i) \leqslant g(S_j)$. Consider finally a vertex $v_l \in S_j$ of probability p_l. Applying [5.8] with $g(S_i)$, $g(S_j)$ and $1 - p_l$ instead of p_1, p_2 and q, respectively, we get:

$$\begin{aligned} 1 - \prod_{v_k \in S_i}(1 - p_k) + 1 - \prod_{v_k \in S_j}(1 - p_k) \geqslant \\ 1 - \prod_{v_k \in S_i \cup \{v_l\}}(1 - p_k) + 1 - \prod_{v_k \in S_j \setminus \{v_l\}}(1 - p_k) \end{aligned} \quad [5.9]$$

Denote by $\hat{S}$ the color of C minimizing quantity $g(S_i) = \prod_{v_k \in S_i}(1 - p_k)$, over any other color S_i of C. Iterating [5.9] for any vertex outside $\hat{S}$ by moving it into $\hat{S}$ (obtaining so a possibly infeasible coloring), we get the inequality claimed.

The proof of [5.9] also shows the following interesting "local optimality" corollary that will be broadly used later.

COROLLARY 5.1.– Consider a coloring $C = (S_1, \ldots S_k)$ of a graph G and two colors S_i and S_j of C. Set $f(S_i) = 1 - \prod_{v_k \in S_i}(1 - p_k)$ and $f(S_j) = 1 - \prod_{v_k \in S_j}(1 - p_k)$ and assume that $f(S_i) \geqslant f(S_j)$. Then, emptying S_j by moving its vertices into S_i produces a (possibly unfeasible) coloring C' such that $E(G, C') \leqslant E(G, C)$.

The proof of the bounds claimed in [5.4] is given in section 5.9.2.

5.2. Basic properties of probabilistic coloring

5.2.1. *Properties under non-identical vertex-probabilities*

We give in this section some general properties about probabilistic coloring, which we use later in order to achieve our results. In what follows, given an *a priori* k-coloring $C = (S_1, \ldots, S_k)$ we will sometimes set, for simplicity, $f(C) = E(G, C)$, where $E(G, C)$ is given by [5.2], and, for $i = 1, \ldots, k$, $f(S_i) = 1 - \prod_{v_j \in S_i}(1 - p_j)$.

PROPERTY 5.1.– Let $C = (S_1, \ldots, S_k)$ be a k-coloring and assume that colors are numbered so that $f(S_i) \leqslant f(S_{i+1})$, $i = 1, \ldots, k-1$. Consider a vertex x (of probability p_x) colored with S_i and a vertex y (of probability p_y) colored with S_j, $j > i$, such that $p_x \geqslant p_y$. If swapping colors of x and y leads to a new feasible coloring C', then $f(C') \leqslant f(C)$.

The proof of Property 5.1 can be found in section 5.9.3.

Notice that Corollary 5.1 can be equivalently stated as follows.

PROPERTY 5.2.– (Equivalent statement of Corollary 5.1) Let $C = (S_1, \ldots, S_k)$ be a k-coloring and assume that colors are numbered so that $f(S_i) \leqslant f(S_{i+1})$, $i = 1, \ldots, k-1$. Consider a vertex x colored with color S_i. If it is feasible to color x with another color S_j, $j > i$, (by keeping colors of the other vertices unchanged), then the new feasible coloring C' verifies $f(C') \leqslant f(C)$.

PROPERTY 5.3.– In any graph of maximum degree Δ, the optimal solution of PROBABILISTIC MIN COLORING contains at most $\Delta + 1$ colors.

Proof. If an optimal coloring C uses $\Delta + k$ colors, $k > 0$, then, by emptying the least-value color (which is always possible as there are at least $\Delta + 1$ colors) and due to Property 5.2, we achieve a $\Delta + 1$-coloring feasible for G with value better (smaller) than the one of C. ∎

5.2.2. *Properties under identical vertex-probabilities*

Properties shown up to this point work for any graph and for any vertex-probability system. Let us now focus on the case of identical vertex-probabilities. It should be noted first that, for this case, Property 5.2 has a natural counterpart expressed as follows.

PROPERTY 5.4.– Let $C = (S_1, \ldots, S_k)$ be a k-coloring and assume that colors are numbered so that $|S_i| \leqslant |S_{i+1}|$, $i = 1, \ldots, k-1$. If it is feasible to inflate a color S_j by "emptying" another color S_i with $i < j$, then the new coloring C', thus created, satisfies $f(C') \leqslant f(C)$.

Proof. Simply note that if $|S_i| \leqslant |S_j|$, then $f(S_i) \leqslant f(S_j)$ and apply the same proof as for Property 5.1. ■

Since, in the proof of Property 5.4, only the cardinalities of the colors intervene, the following corollary-property consequently holds.

PROPERTY 5.5.– Let $C = (S_1, \ldots, S_k)$ be a k-coloring and assume that colors are numbered so that $|S_i| \leqslant |S_{i+1}|$, $i = 1, \ldots, k-1$. Consider two colors S_i and S_j, $i < j$, and a vertex-set $X \subset S_j$ such that $|S_i| + |X| \geqslant |S_j|$. Consider (possibly unfeasible) coloring $C' = (S_1, \ldots, S_i \cup X, \ldots, S_j \setminus X, \ldots, S_k)$. Then, $f(C') \leqslant f(C)$.

We note that Property 5.1, 5.2 or 5.4 describes a process of achieving "locally optima" colorings by local swaps of vertices aiming to "reinforce" the heavier (larger, in the case of identical probabilities) colors. In the sequel, a coloring for which no swaps as the ones described in the statements of Property 5.1, 5.2 or 5.4 are possible, will be called *locally optimal*. Obviously, for a non locally optimal coloring C, there exists a coloring C', better than C, obtained as described in Property 5.1, 5.2 or 5.4. Hence, the following Proposition immediately holds.

PROPOSITION 5.1.– *For any non locally optimal coloring, there exists a locally optimal one dominating it.*

5.3. PROBABILISTIC MIN COLORING in general graphs

5.3.1. *The complexity of probabilistic coloring*

Revisit [5.2] and observe that if we assume $p_i = 1$, $i = 1, \ldots, n$, then PROBABILISTIC MIN COLORING becomes the classical MIN COLORING problem, since, in this case, the weight of any color becomes 1 and what has to be minimized is the number of independent sets in the coloring. This observation immediately deduces the **NP**-hardness of PROBABILISTIC MIN COLORING.

THEOREM 5.1.– PROBABILISTIC MIN COLORING *is* ***NP****-hard.*

5.3.2. *Approximation*

5.3.2.1. *The main result*

In this section, we devise and analyze an approximation algorithm for general PROBABILISTIC MIN COLORING. We first note the following.

REMARK 5.1.– Assume that at least $p_{\max}$ is a fixed constant and denote it by p. Then, denoting, as previously, by $C^* = (S_1^*, \ldots, S_{k^*}^*)$ an optimal *a priori* solution for

PROBABILISTIC MIN COLORING, one gets using [5.48], in the proof of [5.4] in section 5.9.2: $E(G, C^*) \geqslant 1 - \exp\{-p_{\max}\} = 1 - \exp\{-p\}$, which is a fixed constant since p is supposed fixed. If one applies the polynomial algorithm of [LOV 75] computing, in any connected graph G of maximum degree $\Delta \geqslant 3$ (that does not contain a $K_{\Delta+1}$, i.e., a complete graph on $\Delta + 1$ vertices, as induced subgraph) , a coloring with at most Δ colors, this algorithm, when used for PROBABILISTIC MIN COLORING, guarantees (using [5.4]) an approximation ratio of $O(\Delta)$.

Let us first deal with vertex-probabilities such that $p_{\min} \geqslant t$, for some t. Then, the following auxiliary lemma holds.

LEMMA 5.1.– *Assume a graph of order n with vertex-probabilities verifying $p_{\min} \geqslant t$. If minimum coloring is polynomially approximable within approximation ratio ρ, then* PROBABILISTIC MIN COLORING *is approximable in polynomial time within ratio ρ/t.*

Proof. From [5.4], denoting by $C^* = (S_1^*, \ldots, S_{k^*}^*)$ an optimal *a priori* solution for PROBABILISTIC MIN COLORING, we get:

$$E\left(G, C^*\right) \geqslant k^* p_{\min} \geqslant k^* t \qquad [5.10]$$

Consider a ρ-approximation algorithm A computing a feasible coloring $\hat{C}$ for G (by not taking probabilities into account), and set $\hat{C} = (\hat{S}_1, \ldots, \hat{S}_{\hat{k}})$. Then, the functional $E(G, \hat{C})$ for $\hat{C}$, in other words, the objective value of $\hat{C}$ for PROBABILISTIC MIN COLORING, is, by [5.4]:

$$E\left(G, \hat{C}\right) = \sum_{j=1}^{\hat{k}} \left(1 - \prod_{v_i \in \hat{S}_j} \left(1 - p_i\right)\right) \leqslant \hat{k}$$

By hypothesis, $\hat{k}/\chi(G) \leqslant \rho$ (where $\chi(G)$ denotes the chromatic number of the graph G, see Appendix A.2); furthermore, C^* being a feasible coloring, $k^* \geqslant \chi(G)$. Therefore, $\hat{k}/k^* \leqslant \rho$ and the approximation ratio of A for PROBABILISTIC MIN COLORING is, taking [5.10] into account, $E(G, \hat{C})/E(G, C^*) \leqslant \rho/t$. ∎

COROLLARY 5.2.– If $p_{\min}$ is a fixed constant, then PROBABILISTIC MIN COLORING is approximable in polynomial time within ratio $O(\rho)$.

Consider now a graph G, fix two vertex-probabilities p_0 and p', $p_0 < p'$, assume that a ρ-approximation polynomial time algorithm A for minimum coloring exists, and run the following algorithm, called PCOLOR:

1) partition the vertices of G into three subsets: the first, V_{sp} including the vertices with "small" probabilities, i.e., at most p_0, the second, V_{ip}, including the ones with "intermediate" probabilities, i.e., greater than p_0 and at most p', and the third, V_{lp}, including the vertices with "large" probabilities, i.e., greater than p';

2) feasibly color vertices of $G[V_{\text{sp}}]$ and $G[V_{\text{ip}}]$ using a proper set of colors for any subgraph;

3) run A in $G[V_{\text{lp}}]$;

4) take the union of colors computed in Steps 2 and 3 as solution for G.

For simplicity we fix $p_0 = 1/n$. This, as we will see, has no important impact on the approximation ratio concluded; p' will be fixed later.

The following lemmata deal with the approximation ratios of the algorithm above in $G[V_{\text{sp}}]$, $G[V_{\text{ip}}]$ and $G[V_{\text{lp}}]$, respectively. As previously, set $C^* = (S_1^*, \ldots, S_{k^*}^*)$ an optimal *a priori* solution and by $\hat{C} = (\hat{S}_1, \ldots, \hat{S}_{\hat{k}})$ the approximate coloring computed in Step 4. In the proof of the three lemmata just below, C^* and $\hat{C}$ will deal with $G[V_{\text{sp}}]$, $G[V_{\text{ip}}]$ and $G[V_{\text{lp}}]$, respectively.

LEMMA 5.2.– *(The ratio in* $\mathbf{G[V_{sp}]}$*) Any feasible polynomial time approximation algorithm for* PROBABILISTIC MIN COLORING *achieves in* $G[V_{\text{sp}}]$ *approximation ratio bounded above by 2.*

Proof. Denote by n_{sp} the order of $G[V_{\text{sp}}]$. Using [5.3] for $\hat{C}$ and C^*, we get: $E(G[V_{\text{sp}}], \hat{C}) \leqslant \sum_{i=1}^{n_{\text{sp}}} p_i$ and $E(G[V_{\text{sp}}], C^*) \geqslant \sum_{i=1}^{n_{\text{sp}}} p_i - \sum_{i=1}^{n_{\text{sp}}} \sum_{j=i+1}^{n_{\text{sp}}} p_i p_j$. Combining them we get:

$$\frac{E\left(G\left[V_{\text{sp}}\right], C^*\right)}{E\left(G\left[V_{\text{sp}}\right], \hat{C}\right)} \geqslant 1 - \frac{\sum_{i=1}^{n_{\text{sp}}} \sum_{j=i+1}^{n_{\text{sp}}} p_i p_j}{\sum_{i=1}^{n_{\text{sp}}} p_i} = 1 - \frac{\left(\sum_{i=1}^{n_{\text{sp}}} p_i\right)^2 - \sum_{i=1}^{n_{\text{sp}}} p_i^2}{2 \sum_{i=1}^{n_{\text{sp}}} p_i}$$

$$\geqslant 1 - \frac{\sum_{i=1}^{n_{\text{sp}}} p_i}{2} + \frac{\sum_{i=1}^{n_{\text{sp}}} p_i^2}{2 \sum_{i=1}^{n_{\text{sp}}} p_i} \geqslant 1 - \frac{\sum_{i=1}^{n_{\text{sp}}} p_i}{2} \qquad [5.11]$$

Since p_is are assumed to be smaller than $1/n$ and $n_{\text{sp}} \leqslant n$, the right-hand side of [5.11] is at least $1 - n_{\text{sp}}/2n \geqslant 1/2$. Observe now that the approximation ratio of a coloring algorithm in $G[V_{\text{sp}}]$ is $E(G[V_{\text{sp}}], \hat{C})/E(G[V_{\text{sp}}], C^*)$ which is smaller than, or equal to, 2 and the proof of Lemma 5.2 is complete. ■

LEMMA 5.3.– *(The ratio in* $G[V_{ip}]$*) Any feasible polynomial time approximation algorithm for* PROBABILISTIC MIN COLORING *achieves in* $G[V_{ip}]$ *approximation ratio bounded above by* $O(np')$.

Proof. We deal here with a subgraph of G for which, for any vertex v_i, $p_i > p_0 = 1/n$. Obviously, $\prod_{v_i \in S_j^*}(1 - p_i) \leqslant (1 - (1/n))^{|S_j^*|}$ and, consequently:

$$1 - \prod_{v_i \in S_j^*}(1 - p_i) \geqslant 1 - \left(1 - \frac{1}{n}\right)^{|S_j^*|} \geqslant \frac{|S_j^*|}{n} - \frac{|S_j^*|\left(|S_j^*| - 1\right)}{2n^2}$$

where the last inequality is an easy application of the left-hand side of [5.5] with $p_i = 1/n$ for any vertex v_i. Furthermore:

$$\begin{aligned} \frac{|S_j^*|}{n} - \frac{|S_j^*|\left(|S_j^*| - 1\right)}{2n^2} &= \frac{|S_j^*|}{n}\left(1 - \frac{|S_j^*| - 1}{2n}\right) \\ &\geqslant \frac{|S_j^*|}{n}\frac{n+1}{2n} \geqslant \frac{|S_j^*|}{2n} \end{aligned} \quad [5.12]$$

Summing inequality [5.12] for $j = 1, \ldots, k^*$, we get $E(G[V_{ip}], C^*) \geqslant n_{ip}/2n$, where n_{ip} is the order of $G[V_{ip}]$.

On the other hand, using [5.4] we immediately get $E(G[V_{ip}], \hat{C}) \leqslant n_{ip}p'$.

Consequently, using the bounds for $E(G[V_{ip}], C^*)$ and $E(G[V_{ip}], \hat{C})$ provided, we get $E(G[V_{ip}], \hat{C})/E(G[V_{ip}], C^*) \leqslant 2np' = O(np')$, and the proof of Lemma 5.3 is complete. ∎

Finally remark that Lemma 5.1 induces the following lemma dealing with the approximation of PROBABILISTIC MIN COLORING in $G[V_{lp}]$.

LEMMA 5.4.– ***(The ratio in*** $G[V_{lp}]$***)*** *Assuming that* `A` *achieves approximation ratio* ρ *for minimum coloring problem, algorithm* `PCOLOR`*, when running in* $G[V_{lp}]$*, achieves approximation ratio bounded above by* ρ/p' *for* PROBABILISTIC MIN COLORING.

We are going now to use Lemmata 5.2, 5.3 and 5.4 in order to complete the analysis of the overall algorithm.

THEOREM 5.2.– *On the hypothesis that* `A` *guarantees approximation ratio* ρ *for minimum coloring,* `PCOLOR` *approximately solves in polynomial time* PROBABILISTIC MIN COLORING *within ratio* $O(\sqrt{\rho n})$.

Proof. In what follows, denote by C^* an optimal *a priori* coloring for PROBABILISTIC MIN COLORING in G, by $C^*[V_{\text{sp}}]$, $C^*[V_{\text{ip}}]$ and $C^*[V_{\text{lp}}]$ the solutions induced by C^* in $G[V_{\text{sp}}]$, $G[V_{\text{ip}}]$ and $G[V_{\text{lp}}]$, respectively, and by C^*_{sp}, C^*_{ip} and C^*_{lp}, $\hat{C}_{\text{sp}}$, $\hat{C}_{\text{ip}}$ and $\hat{C}_{\text{lp}}$ the optimal and approximated *a priori* solutions in $G[V_{\text{sp}}]$, $G[V_{\text{ip}}]$ and $G[V_{\text{lp}}]$, respectively.

We prove that, for any $x \in \{\text{sp}, \text{ip}, \text{lp}\}$:

$$E(G, C^*) \geqslant E(G[V_x], C^*[V_x]) \geqslant E(G[V_x], C^*_x)$$

Note that $C^*[V_x]$ is a particular feasible solution for $G[V_x]$; hence:

$$E(G[V_x], C^*[V_x]) \geqslant E(G[V_x], C^*_x)$$

In order to prove the first inequality, fix an x and consider a color, say S^*_j of C^* Then, the contribution of S^*_j in $C^*[V_x]$ is:

$$1 - \prod_{v_i \in S^*_j \cap V_x} (1 - p_i) \leqslant 1 - \prod_{v_i \in S^*_j} (1 - p_i)$$

which is its contribution in C^*. Iterating this argument for all the colors in $C^*[V_x]$, the claim follows.

Recall finally that the algorithm colors the vertices of $G[V_x]$, $x \in \{\text{sp}, \text{ip}, \text{lp}\}$ with a distinct set of colors and the *a priori* solution $\hat{C}$ finally provided is the union of these sets. Consequently:

$$E\left(G, \hat{C}\right) = E\left(G[V_{\text{sp}}], \hat{C}_{\text{sp}}\right) + E\left(G[V_{\text{ip}}], \hat{C}_{\text{ip}}\right) + E\left(G[V_{\text{lp}}], \hat{C}_{\text{lp}}\right)$$

Note that $E(G, C^*)$ is at least as large as any of $E(G[V_x], C^*_x)$, $x \in \{\text{sp}, \text{ip}, \text{lp}\}$. Hence, one immediately deduces that the overall ratio of the algorithm in G is at most the sum of the ratios proved by Lemmata 5.2, 5.3 and 5.4, i.e., at most $O(2 + np' + (\rho/p'))$.

Remark that the ratio claimed in Lemma 5.3 is increasing with p', while the one in Lemma 5.4 is decreasing with p'. Equality of expressions np' and ρ/p' holds for $p' = \sqrt{\rho/n}$. In this case the value of the ratio obtained is $O(\sqrt{\rho n})$, and the proof of the theorem is now complete. ∎

COROLLARY 5.3.– Using for A the $O(n \log^2 \log n / \log^3 n)$-approximation algorithm of [HAL 93], the approximation ratio achieved by PCOLOR is $O(n \log\log n / \log^{3/2} n)$.

Moreover, taking into account that any induced subgraph of a k-colorable graph is k-colorable, the following corollary holds also.

COROLLARY 5.4.– If A stands for the $O(n^{1-(3/(k+1))}\log n)$-approximation algorithm of [KAR 98], then the approximation ratio achieved by PCOLOR in k-colorable graphs is $O(n^{1-(3/(2k+2))}\sqrt{\log n})$.

5.3.2.2. *Further approximation results*

We first note that another easy corollary of Theorem 5.2 is that when the minimum coloring problem is polynomial, PROBABILISTIC MIN COLORING is approximable within approximation ratio $O(\sqrt{n})$.

On the other hand, it is easy to see that if there exists an algorithm producing a k-coloring, then it achieves approximation ratio bounded above by k. Indeed, if such an algorithm produces a coloring $C = (S_1, \ldots, S_k)$ in a graph $G(V, E)$, then $E(G, C) = \sum_{j=1}^{k}(1 - \prod_{v_i \in S_j}(1 - p_i)) \leqslant \sum_{j=1}^{k}(1 - \prod_{v_i \in V}(1 - p_i)) = k(1 - \prod_{v_i \in V}(1 - p_i))$. Using the leftmost lower bound of [5.3] for $E(G, C^*)$, where C^* denotes an optimal probabilistic coloring of G, the bound claimed is immediately concluded.

Let us focus ourselves on graph-classes for which the (deterministic) minimum coloring problem is polynomial in both the input graph itself and any subgraph of it. Then the following corollary holds.

COROLLARY 5.5.– PROBABILISTIC MIN COLORING is approximable within approximation ratio k in graphs of chromatic number k where the minimum coloring problem is polynomial in the input graph and in any of its induced subgraphs. In particular, if k is fixed, then PROBABILISTIC MIN COLORING $\in$ **APX**, the class of problems approximable within constant ratio.

Let us further restrict ourselves to graphs where *not only the computation of a minimum coloring but also the computation of a maximum-weight independent set is polynomial.* For this case, consider the following algorithm:

1) charge any vertex $v_i \in V$ with weight $\log((1 - p_i)^{-1})$;

2) compute a maximum-weight independent set S^* and color its vertices with the same color;

3) solve the minimum coloring in $G[V \setminus S^*]$;

4) output the coloring $C = (S^*, S_1, \ldots, S_k)$ thus computed (note that S_k may be empty).

With simple algebra, for any $S \subseteq V$:

$$1 - \prod_{v_i \in S}(1 - p_i) = 1 - \exp\left\{-\sum_{v_i \in S}\log\left(\frac{1}{(1 - p_i)}\right)\right\}$$

Note also that function $f(x) = -e^{-x}$, is increasing with x, for any x. So, the independent set maximizing $\sum_{v_i \in S} \log(1-p_i)^{-1}$ is exactly the one maximizing $1 - \prod_{v_i \in S}(1-p_i)$. Hence, the independent set S^*, computed in Step 1 of the algorithm above, represents the maximum-value feasible color of C. Considering that colors in C are ranged in decreasing value order:

$$E(G, C) = \left(1 - \prod_{v_i \in S^*} (1-p_i)\right) + \sum_{j=1}^{k} \left(1 - \prod_{v_i \in S_j} (1-p_i)\right) \qquad [5.13]$$

Set, for simplicity, $\alpha = (1 - \prod_{v_i \in S^*}(1-p_i))$ and, for $j = 1, \ldots, k$, $\beta_j = (1 - \prod_{v_i \in S_j}(1-p_i))$ and suppose without loss of generality that $\beta_1 \geqslant \beta_2 \geqslant \ldots \geqslant \beta_k$. Then, [5.13] becomes:

$$E(G, C) = \alpha + \sum_{j=1}^{k} \beta_j \qquad [5.14]$$

In the same spirit as in Corollary 5.1, one can see that, starting from some coloring, one can obtain a better-value (possibly unfeasible) coloring not only by strictly inflating the heaviest color of it, but also by swapping vertices in such a way that the final heaviest color corresponds to an independent set S maximizing $1 - \prod_{v_i \in S}(1-p_i)$ (or, equivalently, $\sum_{v_i \in S} \log(1-p_i)^{-1}$). So, the value of an optimal solution $C^* = (S_1^*, \ldots, S_{k^*}^*)$ (considering that these colors are also ranged in decreasing value order) is greater than or equal to $(1 - \prod_{v_i \in S^*}(1-p_i)) + \sum_{j=2}^{k^*}(1 - \prod_{v_i \in S_j^*}(1-p_i))$. Then, using Corollary 5.1 in the graph $G[V \setminus S^*]$:

$$E(G, C^*) \geqslant \left(1 - \prod_{v_i \in S^*} (1-p_i)\right) + \left(1 - \prod_{v_i \in V \setminus S^*} (1-p_i)\right) \qquad [5.15]$$

Set $\gamma = (1 - \prod_{v_i \in V \setminus S^*}(1-p_i))$. Then, [5.15] becomes:

$$E(G, C^*) \geqslant \alpha + \gamma \qquad [5.16]$$

Note now that since α represents the weight of the maximum-weight independent set, then by the discussion just after the outline of the algorithm: $\alpha \geqslant \beta_j$, for any

$j = 1, \ldots, k$; hence, $E(G, C) \leqslant (k+1)\alpha$. Also, since the product in any of β_j is composed by a subset of terms in the product in γ, $\beta_j \leqslant \gamma$, for any $j = 1, \ldots, k$; henceforth, $E(G, C) \leqslant \alpha + k\gamma$. Taking all this into account, [5.14] becomes:

$$E(G, C) \leqslant \min\{(k+1)\alpha, \alpha + k\gamma\} \qquad [5.17]$$

If $(k+1)\alpha \leqslant \alpha + k\gamma$, then $\alpha \leqslant \gamma$ and combination of [5.17] and [5.16] results in $E(G, C)/E(G, C^*) \leqslant (k+1)/2$. On the other hand, if $(k+1)\alpha \geqslant \alpha + k\gamma$, then $\alpha \geqslant \gamma$ and, taking into account that function $f(\alpha) = (\alpha + k\gamma)/(\alpha + \gamma)$ decreases with α, combination of [5.17] and [5.16] results, once more, in $E(G, C)/E(G, C^*) \leqslant (k+1)/2$.

COROLLARY 5.6.– PROBABILISTIC MIN COLORING is approximable within approximation ratio $(k+1)/2$ in graphs of chromatic number k where the minimum coloring problem and the maximum-weight independent set problem are both polynomial.

Corollary 5.6 has the following interesting instantiation when dealing with bipartite graphs (where $k = 2$).

COROLLARY 5.7.– PROBABILISTIC MIN COLORING is approximable within approximation ratio $3/2$ in bipartite graphs.

5.4. PROBABILISTIC MIN COLORING in bipartite graphs

We denote by $B(V, U, E)$ a connected bipartite graph with bipartition V and U and edge-set E. We first make the following emphasized preliminary observation: *in any connected bipartite graph, the bipartition (bicoloring) of its vertices is unique.* This unique 2-coloring is not always the best *a priori* solution for PROBABILISTIC MIN COLORING in a bipartite graph as it is shown in Figure 5.3, where the functional of the 3-coloring consisting of taking v_1, v_2, v_4 and u_2 in the first color, u_1 and u_3 in the second color and v_3 in the third color is equal to 1.3896 and better than the one induced by the 2-coloring (V, U), equal to 1.8364.

5.4.1. *A basic property*

This section completes the discussion in section 5.2 by establishing a further property for probabilistic coloring in bipartite graphs, under identical vertex probabilities. We first note that for "trivial" families of bipartite graphs, as graphs isomorphic to a perfect matching, or to an independent set (i.e., a collection of isolated vertices), PROBABILISTIC MIN COLORING is polynomial, under any system of vertex-probabilities. In fact, for the former case, the optimal solution is given by a 2-coloring

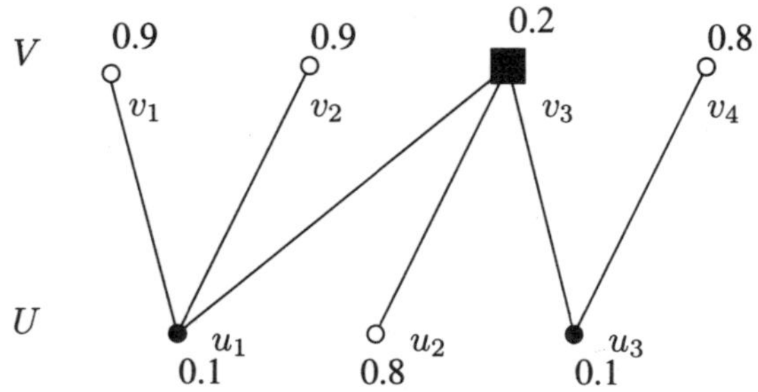

Figure 5.3. *A bipartite graph $B(V, U, E)$ where the 2-coloring is not the best-functional a priori solution*

where for each pair of matched vertices, the one with largest probability is assigned to the first color, while the other one is assigned to the second color. For the latter case, trivially, the 1-coloring is optimal.

Observe also that the cases of vertex-probability 0 or 1 are trivial: for the former, any *a priori* solution has value 0; for the latter, PROBABILISTIC MIN COLORING coincides with the classical (deterministic) coloring problem where the (unique) 2-coloring is the best one.

Consider a bipartite graph $B(V, U, E)$ and, without loss of generality, assume $|V| \geqslant |U|$. Also, denote by $\alpha(B)$ the cardinality of a maximum independent set of B. Then the following property holds.

PROPERTY 5.6.– If $\alpha(B) = |V|$, then 2-coloring $C = (V, U)$ is optimal.

Proof. Suppose *a contrario* that C is not optimal; then, the optimal coloring C' uses exactly $k \geqslant 3$ colors and its largest cardinality color S'_1 has cardinality β. Consider the following exhaustive two cases:

$\boldsymbol{\alpha(B) = \beta}$: then, it is sufficient to aggregate all the vertices not belonging to S'_1 into another color, say S'_2; this would lead to a – possibly unfeasible – solution C'' which improves upon C' (due to Proposition 5.1) and whose value coincides with the value of C;

$\boldsymbol{\alpha(B) > \beta}$: assume that one adds to color S'_1 exactly $\alpha(B) - \beta$ vertices from the other colors neglecting possible unfeasibilities; the resulting solution C'' dominates C' (due to Proposition 5.1); but then, the largest cardinality color S''_1 has in solution C'' exactly $\alpha(B)$ vertices; hence, as for case $\alpha(B) = \beta$, the 2-coloring C is feasible, and dominates both C'' and C'. ∎

5.4.2. *General bipartite graphs*

Let us first note that the complexity of PROBABILISTIC MIN COLORING in general bipartite graphs still remains open. However, we strongly argue that the problem is **NP**-hard.

In what follows, we first give an easy result showing that the presumably hard cases for PROBABILISTIC MIN COLORING are the ones where vertex-probabilities are "small". Consider a bipartite graph $B(V, U, E)$ and denote by $p_{\min}$ its smallest vertex-probability. The following proposition, proved in section 5.9.4, holds.

PROPOSITION 5.2.– *If* $p_{\min} \geqslant 0.5$, *then 2-coloring* $C = (V, U)$ *is optimal for* B.

When vertex-probabilities are generally and typically smaller than 0.5, the situation completely changes with respect the result of Proposition 5.2. Indeed, in this case, it is possible to provide instances, even with identical vertex-probabilities, where the 2-coloring does not provide the optimal solution. For instance, consider the tree T of Figure 5.4, where vertex 1 (the tree's root) is linked to vertices $n+1, \dots, 2n$ and vertex $2n$ is linked to vertices $1, \dots, n$.

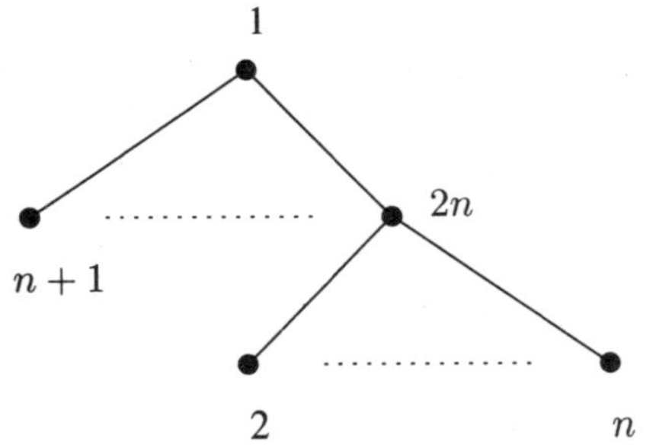

Figure 5.4. *A tree with a 3-coloring of better value than the one of its 2-coloring*

Assume that vertex-probabilities of the vertices of T are all equal to $p \ll 0.5$. Then, the 2-coloring $\{\{1, \dots, n\}, \{n+1, \dots, 2n\}\}$ has value $f_2 = 2(1 - (1-p)^n)$, while the 3-coloring $\{\{1\}, \{2, \dots, 2n-1\}, \{2n\}\}$ has value $f_3 = 2(1 - (1-p)) + (1 - (1-p)^{2n-2})$. For p small enough and n large enough, we have $f_2 \approx 2$ and $f_3 \approx 1$.

The example of Figure 5.4 generalizes the counter-example of [MUR 03], dealing only with bipartite graphs, and shows that not only in general bipartite graphs but even in trees (that are restricted cases of bipartite graphs), the obvious 2-coloring is not always the optimal solution of PROBABILISTIC MIN COLORING.

Dealing with distinct-vertex-probabilities, the following proposition holds.

PROPOSITION 5.3.– *In any bipartite graph B(V,U,E), its 2-coloring $C = (V, U)$ achieves approximation ratio bounded by 2. This bound is tight even on paths.*

Proof. Consider a bipartite graph $B(V, U, E)$. A trivial lower bound on the optimal solution cost (due to Property 5.1) is given by the unfeasible 1-coloring $V \cup U$ with all the vertices having the same color. Hence, denoting by C^* an optimal coloring of B, we have:

$$f(V \cup U) \leqslant f(C^*) \qquad [5.18]$$

Assume that $f(V) \leqslant f(U)$. Then, since $U \subseteq V \cup U$, $f(U) \leqslant f(V \cup U)$. Therefore, using [5.18] $f(C) = f(V) + f(U) \leqslant 2f(U) \leqslant 2f(V \cup U) \leqslant 2f(C^*)$, QED.

Figure 5.5. *Ratio 2 is tight for the 2-coloring of a bipartite graph*

For tightness, consider the 4-vertex path of Figure 5.5. The 2-coloring has value $2 - 2\epsilon + 2\epsilon^2$, while the 3-coloring $\{1, 4\}, \{2\}, \{3\}$ has value $1 + 2\epsilon - \epsilon^2$. For $\epsilon \to 0$, the latter is the optimal solution and the approximation ratio of the two coloring tends to 2. ∎

Consider now the following algorithm, denoted by `3-WEIGHTED_COLOR` in what follows:

1) compute and store the natural 2-coloring $C_0 = (V, U)$;

2) compute a maximum weighted independent set S of B;

3) output the best coloring among C_0 and $C_1 = (S, V \setminus S, U \setminus S)$.

Computation of a maximum-weight independent set can be performed in polynomial time in bipartite graphs ([BOU 84]); so, 3-WEIGHTED_COLOR is polynomial.

PROPOSITION 5.4.– *Algorithm 3-WEIGHTED_COLOR achieves approximation ratio bounded above by 8/7 in bipartite graphs.*

Proof. Consider an optimal solution $C^* = (S_1^*, S_2^*, \ldots S_k^*)$ and assume that $f(S_1^*) \geqslant f(S_2^*) \geqslant \ldots \geqslant f(S_k^*)$. Based upon Property 5.2, the worst case for C_0 is reached when it is completely balanced, i.e., when $f(V) = f(U)$. Using the fact that, given two rational numbers with constant product, their sum is minimized when they are equal, we get after some very easy algebra:

$$f(C_0) = f(V) + f(U) \leqslant 2\left(1 - \prod_{v_i \in V \cup U} (1 - p_i)^{\frac{1}{2}}\right) \quad [5.19]$$

By exactly the same reasoning, the worst case for C_1 is reached when $f(V \setminus S) = f(U \setminus S)$, namely when:

$$\begin{aligned} f(C_1) &= f(S) + f(V \setminus S) + f(U \setminus S) \\ &\leqslant f(S) + 2\left(1 - \prod_{v_i \in (V \cup U)\setminus S} (1 - p_i)^{\frac{1}{2}}\right) \\ &\leqslant 1 - \prod_{v_i \in S} (1 - p_i) + 2\left(1 - \prod_{v_i \in (V \cup U)\setminus S} (1 - p_i)^{\frac{1}{2}}\right) \end{aligned} \quad [5.20]$$

Recall that $f(S_1^*) \leqslant f(S)$ and, henceforth, due to Property 5.2:

$$\begin{aligned} &f(C^*) \geqslant f(S) + f((V \cup U) \setminus S) \\ &= 1 - \prod_{v_i \in S} (1 - p_i) + 1 - \prod_{v_i \in (V \cup U)\setminus S} (1 - p_i) \end{aligned} \quad [5.21]$$

Setting $\beta = \prod_{v_i \in S} (1 - p_i)^{\frac{1}{2}}$; $\alpha = \prod_{v_i \in (V \cup U)\setminus S} (1 - p_i)^{\frac{1}{2}}$ and using [5.19], [5.20] and [5.21], we get for the approximation ratio ρ the following expression:

$$\begin{aligned} \rho(B) &= \min\left\{\frac{f(C_0)}{f(C^*)}, \frac{f(C_1)}{f(C^*)}\right\} \\ &\leqslant \min\left\{\frac{2(1 - \alpha\beta)}{2 - \alpha^2 - \beta^2}, \frac{3 - \beta^2 - 2\alpha}{2 - \alpha^2 - \beta^2}\right\} \end{aligned} \quad [5.22]$$

We now show that function $f_1(x) = 2(1 - \beta x)/(2 - x^2 - \beta^2)$ increases with x in $[\beta, 1)$, while function $f_2(x) = (3 - \beta^2 - 2x)/(2 - x^2 - \beta^2)$ decreases with x in the same interval. Indeed, by elementary algebra, one immediately gets:

$$f_1'(x) = \frac{-2\beta(x-\beta)\left(x - \left(\frac{2-\beta^2}{\beta}\right)\right)}{\left(2 - x^2 - \beta^2\right)^2} \qquad [5.23]$$

$$f_2'(x) = \frac{-2(x-1)\left(x - \left(2 - \beta^2\right)\right)}{\left(2 - x^2 - \beta^2\right)^2} \qquad [5.24]$$

In [5.23], $(2 - \beta^2)/\beta \geqslant 1$; so, $f_1'(x)$ is positive for $x \in [\beta, 1)$ and, consequently f_1 is increasing with x in this interval. On the other hand, in [5.24], since $x < 1$ and $\beta < 1$, $x - 1 \leqslant 0$ and $x - (2 - \beta^2) \leqslant 0$. So, $f_2'(x)$ is negative for $x \in [\beta, 1)$ and, consequently f_2 is decreasing with x in this interval.

In all, quantity $\min\{f_1(\alpha), f_2(\alpha)\}$ achieves its maximum value for α verifying $f_1(\alpha) = f_2(\alpha)$, or when $2(1 - \alpha\beta) = 3 - \beta^2 - 2\alpha$, i.e., when $\alpha = (1 + \beta)/2$. In this case [5.22] becomes (for $\beta \leqslant 1$):

$$\rho(B) \leqslant \frac{2\left(1 - \left(\frac{1+\beta}{2}\right)\beta\right)}{2 - \left(\frac{1+\beta}{2}\right)^2 - \beta^2} = \frac{8 - 4\beta - 4\beta^2}{7 - 2\beta - 5\beta^2} \leqslant \frac{8}{7}$$

and the result claimed is proved. ∎

Notice that there exist arbitrarily large instances in which, if 3-WEIGHTED_COLOR is allowed to arbitrarily choose some maximum independent set, it achieves approximation ratio asymptotically equal to 8/7. For instance, fix an $n \in \mathbb{N}$ and consider the following bipartite graph $B(V, U, E)$ consisting of:

– an independent set S_1 on $2n^2$ vertices; n^2 of them denoted by $v_V^1, \ldots, v_V^{n^2}$ belong to V and the n^2 remaining ones denoted by $v_U^1, \ldots, v_U^{n^2}$ belong to U;

– n paths $P_1, \ldots, P_n$ of size 4 (i.e. on 3 edges); set, for $i = 1, \ldots, n$, $P_i = (r_i^1, r_i^2, r_i^3, r_i^4)$, where $r_i^1, r_i^3 \in V$ and $r_i^2, r_i^4 \in U$; S_1 and the n paths P_i are disjoint;

– two vertices $v \in V$ and $u \in U$; v is linked to all the vertices of U and u to all the vertices of V;

– for any $v_i \in V \cup U$, $p_i = p = \ln 2/n$.

The graph so-constructed is balanced (i.e., $|V| = |U|$) and has size $2n^2 + 4n + 2$. Figure 5.6 shows such a graph for $n = 2$.

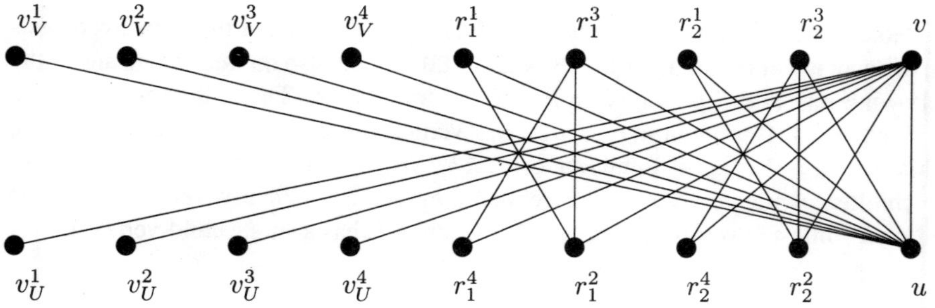

Figure 5.6. *An 8/7 instance for 3-WEIGHTED_COLOR with $n = 2$*

Apply algorithm 3-WEIGHTED_COLOR to the so-constructed graph B. Coloring $C_0 = (V, U)$ has value:

$$f(C_0) = 2\left(1 - (1-p)^{n^2+2n+1}\right) \qquad [5.25]$$

On the other hand, one can see that several maximum-weight independent sets of B exist, each consisting of the $2n^2$ vertices of S_1 plus two vertices per any of the n paths P_i, $i = 1, \ldots, n$. Assume that the maximum-weight independent set computed in Step 2 of algorithm 3-WEIGHTED_COLOR is $S = S_1 \cup_{i=1,\ldots,n} \{r_i^1, r_i^4\}$. In this case, $|S| = 2n^2 + 2n$, and $|V \setminus S| = |U \setminus S| = n + 1$; hence, the value of the coloring $C_1 = (S, V \setminus S, U \setminus S)$ examined in Step 3 has value:

$$f(C_1) = 1 - (1-p)^{2n^2+2n} + 2\left(1 - (1-p)^{n+1}\right) \qquad [5.26]$$

Finally, consider the coloring $\hat{C} = (\hat{S}_1, \hat{S}_2, \hat{S}_3)$ of B where:

– $\hat{S}_1 = S_1 \cup_{i=1,\ldots,n} \{r_i^1, r_i^3\}$;

– $\hat{S}_2 = \{u\} \cup_{i=1,\ldots,n} \{r_i^2, r_i^4\}$;

– $\hat{S}_3 = \{v\}$.

Obviously:

$$f\left(\hat{C}\right) = 1 - (1-p)^{2n^2+2n} + 1 - (1-p)^{2n+1} + p \qquad [5.27]$$

One can easily see that for $n \to \infty$ and for $p = \ln 2/n$, [5.25], [5.26] and [5.27] give respectively: $f(C_0) \to 2$, $f(C_1) \to 2$ and $f(C^*) \leqslant f(\hat{C}) \to 7/4$.

Note that the tightness of the bound 8/7 can be shown (under the same hypothesis on the way it works) for algorithm 3-WEIGHTED_COLOR also on trees by means of the following instance T presented in Figure 5.7 for $n = 2$. There, the root-vertex a_0 of T has $n^2 + 1$ children $a_1, \ldots, a_{n^2}, b_0$. Vertices $\{a_1, \ldots, a_{n^2}\}$ have no children, while vertex b_0 has $n^2 + 1$ children $b_1, \ldots, b_{n^2}, c_0$. Again, vertices $b_1, \ldots, b_{n^2}$ have no children, while vertex c_0 has $2n$ children $c_1, \ldots, c_{2n}$. Finally, vertex c_{2n} has no children while any vertex c_i, with $i = 1, \ldots, 2n - 1$, has a single child-vertex d_i.

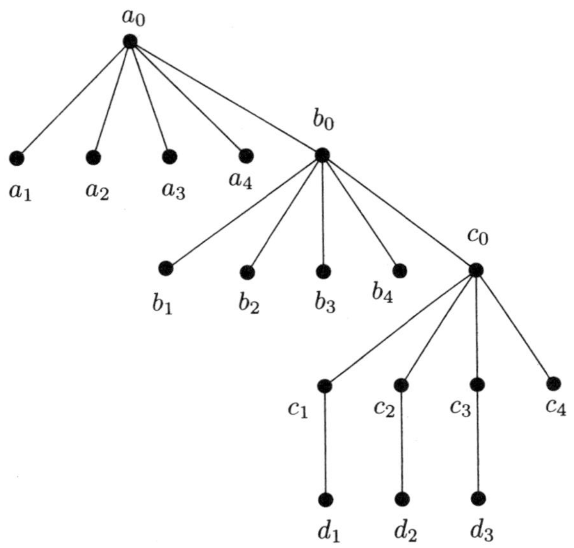

Figure 5.7. *Lower bound 8/7 is attained for 3-WEIGHTED_COLOR even in trees ($n = 2$)*

The tree T so-constructed gives, as in the previous example, a balanced bipartite graph (i.e., $|V| = |U|$) and has size $2n^2 + 4n + 2$. Apply 3-WEIGHTED_COLOR to T and set $C'_0 = (V, U)$. Assume that the maximum independent set computed in Step 2 of the algorithm is $S' = \{a_1, \ldots, a_{n^2}, b_1, \ldots, b_{n^2}, c_{n+1}, \ldots, c_{2n}, d_1, \ldots, d_n\}$. Then the coloring $C' = (S', V \setminus S', U \setminus S')$ is also examined in Step 3. Besides, coloring $\hat{C}' = (\hat{S}'_1, \hat{S}'_2, \hat{S}'_3)$ with $\hat{S}'_1 = \{a_1, \ldots, a_{n^2}, b_1, \ldots, b_{n^2}, c_1, \ldots, c_{2n}\}$, $\hat{S}'_2 = \{a_0, c_0, d_1, \ldots, d_{2n-1}\}$, $\hat{S}'_3 = \{b_0\}$ is the best one. Some easy algebra derives then ratio 8/7 for 3-WEIGHTED_COLOR when running on the considered tree.

Algorithm 3-WEIGHTED_COLOR is a simplified version of the following algorithm, called MASTER-SLAVE[1]:

1) compute and store the natural 2-coloring (V, U);

2) set $B_1(V_1, U_1) = B(V, U)$;

3) set $i = 1$;

4) repeat the following steps until possible:

a) compute some maximum-weight independent set S_i of B_i;

b) set $(V_{i+1}, U_{i+1}) = (V_i \setminus S_i, U_i \setminus S_i)$;

c) compute and store coloring $(S_1, \ldots, S_i, V_{i+1}, U_{i+1})$;

5) compute and store coloring $(S_1, S_2, \ldots)$, where S_is are the independent sets computed during the executions of Step 4a);

6) output C, the best among the colorings computed in Steps 1, 4c) and 5.

This algorithm obviously provides solutions that are at least as good as the ones provided by 3-WEIGHTED_COLOR. Therefore its approximation ratio for PROBABILISTIC MIN COLORING is at most $8/7$. We show that it cannot do better (always as it is, i.e., allowing it to arbitrarily choose the consecutive independent set S_i in Step 4a). Indeed, consider the counter-example after the proof of Proposition 5.4. After computation of S, the surviving graph consists of the vertex-set $\cup_{i=1,\ldots,n}\{r_i^2, r_i^3\} \cup \{v, u\}$. In this graph, the maximum independent set is of size $n+1$ (say the vertices of the surviving subset of V). In other words, colorings C_i computed, for $i \geqslant 2$ by MASTER-SLAVE are the same as coloring C_1 computed by 3-WEIGHTED_COLOR.

Note, however, that the counter-example on trees, presented above, does not work if algorithm MASTER-SLAVE is applied instead of 3-WEIGHTED_COLOR.

5.4.3. *Bipartite complements of bipartite matchings*

We deal in this section with bipartite complements of bipartite matchings, i.e, with bipartite graphs $B(V, U, E)$ with $|V| = |U| = n$ and with $E = E(B_{n,n}) \setminus \{v_i u_i, v_i \in V, u_i \in U, i = 1, \ldots, n\})$, where by $B_{n,n}$ we denote the complete bipartite graph with $|V| = |U| = n$. Such a graph will be denoted by $\bar{M}_{n,n}$ (in Figure 5.8, $\bar{M}_{4,4}$ is illustrated). We will show that, under any vertex-probability system, PROBABILISTIC MIN COLORING is polynomial in this graph-class.

In any graph $\bar{M}_{n,n}$, a color will be called a *horizontal color* if it is a *proper subset* either of V, or of U; a coloring will be called a horizontal coloring if it is composed

1. This kind of algorithms approximately solving a "master" problem (COLORING in this case) by running a subroutine for a maximization "slave" problem (MAX INDEPENDENT SET here) appears for first time in [JOH 74]; appellation "master-slave" for these algorithms is due to [SIM 90].

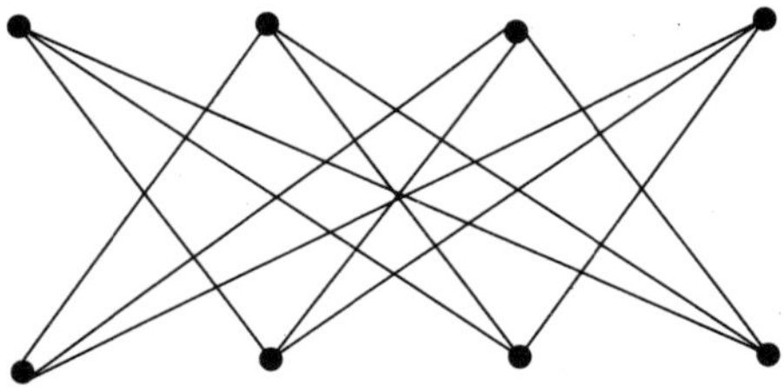

Figure 5.8. *A graph $\bar{M}_{4,4}$*

only by horizontal colors. On the other hand, a color will be called a *vertical color* if it contains vertices from both V and U; a coloring of $\bar{M}_{k,k}$ will be called a vertical coloring if *all* its colors are vertical, otherwise it will be called non-vertical coloring.

LEMMA 5.5.– *The following properties hold for the colorings of $\bar{M}_{k,k}$:*

1) any vertical color of $\bar{M}_{k,k}$ is exclusively of the form $\{v_i, u_i\}$, $i = 1, \ldots, k$;

2) the non-vertical colors of any non-vertical coloring of $\bar{M}_{k,k}$ are horizontal, i.e., there is no coloring of $\bar{M}_{k,k}$ other than with horizontal or vertical colors;

3) for any $i = 1, \ldots, k$, if v_i and u_i belong to two different colors, these colors are horizontal;

4) for any induced subgraph $B'(V', U', E')$ of $\bar{M}_{k,k}$, the functional-value of any horizontal coloring C is greater than, or equal to, the one of the 2-coloring (V', U') of B'; furthermore, if the vertex-probabilities in V' (resp., U') are pairwise distinct, then $E(B', C) > E(B', (V', U'))$.

Proof. Items 1 and 2 are easily deduced from the particular form of $\bar{M}_{k,k}$ implying that independent sets $\{v_i, u_i\}$, $i = 1, \ldots, k$ are all maximal for the inclusion.

Item 3 is concluded by the fact that v_i excludes any vertex of U (other than u_i), while u_i excludes any vertex of V (other than v_i).

Finally, Item 4 is a simple application of Corollary 5.1. Indeed, starting from any horizontal coloring, one can move vertices from some color to a heavier one, until the natural 2-coloring is produced for B'. By the form of a horizontal coloring, all these moves, provided that they are performed inside V, or U, result in new feasible colorings. By Corollary 5.1 (or, equivalently, by Property 5.2), any of them is better (i.e., has smaller value) than the previous one. The second claim of Item 4 can be shown by some easy algebra. ∎

In what follows, we call vertical group the set $R_i = \{v_i, u_i\}$, $i = 1, \ldots, n$. Observe that Properties 1, 2 and 3 of Lemma 5.5 hold independently of the vertex-probability system considered. Observe also that an immediate consequence of the combination of these properties with Corollary 5.1 is the following lemma.

LEMMA 5.6.– *Consider a graph $\bar{M}_{n,n}$. Any optimal coloring of $\bar{M}_{n,n}$ has at most two horizontal colors. If S_1 and S_2 are these colors, then the subgraph of $\bar{M}_{n,n}$ induced by $S_1 \cup S_2$ is a bipartite complement of a bipartite matching.*

The following lemma is the key part for the proof of the result claimed in the beginning of this section. For simplicity, for a group R_i, set $f(R_i) = 1 - (1 - p_{v_i})(1 - p_{u_i})$. Suppose furthermore that, upon a reordering of the groups, $\bar{M}_{n,n}$ is represented in such a way that $f(R_1) \geqslant f(R_2) \geqslant \ldots \geqslant f(R_n)$; in other words "their values diminish from the left to the right".

LEMMA 5.7.– *Consider three vertical groups $R_i = \{v_i, u_i\}$, $R_j = \{v_j, u_j\}$ and $R_k = \{v_k, u_k\}$, such that $f(R_i) \geqslant f(R_j) \geqslant f(R_k)$. Fix an optimal coloring C^* of $\bar{M}_{n,n}$. If colors of v_i, v_j (on the one hand) and of u_i, u_j (on the other hand) are horizontal in C^*, then R_k cannot be a vertical color of C^*.*

Proof. Denote by S_v and S_u the horizontal colors of v_i, v_j and u_i, u_j, respectively and assume, without loss of generality, that $f(S_v) \geqslant f(S_u)$, where, as previously, $f(S)$ denotes the weight of color S. In other words, assume that:

$$\left(1 - p_{v_i}\right)\left(1 - p_{v_j}\right) \leqslant \left(1 - p_{u_i}\right)\left(1 - p_{u_j}\right) \quad [5.28]$$

As it will be understood below, the opposite case (i.e., the one where $(1-p_{v_i})(1-p_{v_j}) \geqslant (1 - p_{u_i})(1 - p_{u_j})$) is completely similar.

By the assumption on the values of R_i, R_j and R_k, the following holds:

$$\left(1 - p_{v_i}\right)\left(1 - p_{u_i}\right) \leqslant \left(1 - p_{v_j}\right)\left(1 - p_{u_j}\right) \leqslant \left(1 - p_{v_k}\right)\left(1 - p_{u_k}\right) \quad [5.29]$$

It suffices to show that $(1-p_{v_i})(1-p_{v_j})\prod_{v_\ell \in S_v \setminus \{v_i, v_j\}}(1-p_\ell) \leqslant (1-p_{v_k})(1-p_{u_k})$, i.e., that color S_v is "heavier" than color R_k. In this case, application of Corollary 5.1 (by "diluting" color R_k to colors S_v and S_u, in this order) allows us to improve solution. Since $(1-p_{v_i})(1-p_{v_j})\prod_{v_\ell \in S_v \setminus \{v_i, v_j\}}(1-p_\ell) \leqslant (1-p_{v_i})(1-p_{v_j})$, it suffices to show that $(1-p_{v_i})(1-p_{v_j}) \leqslant (1-p_{v_k})(1-p_{u_k})$. By [5.29] one immediately gets:

$$\left(1 - p_{v_i}\right)\left(1 - p_{u_i}\right)\left(1 - p_{v_j}\right)\left(1 - p_{u_j}\right) \leqslant \left(1 - p_{v_k}\right)^2\left(1 - p_{u_k}\right)^2 \quad [5.30]$$

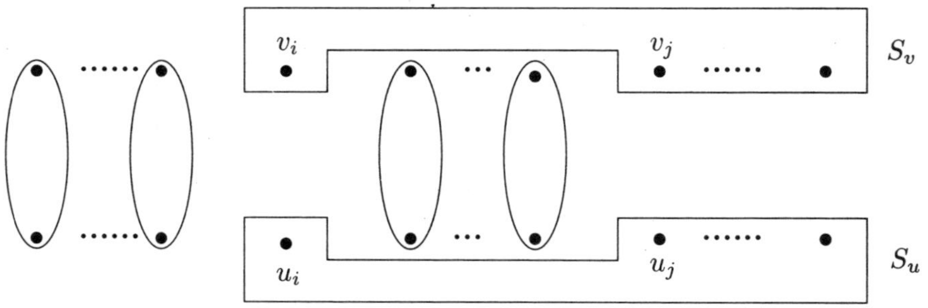

Figure 5.9. *The coloring implied by Lemma 5.7*

On the other hand, combination of [5.28] and [5.30] easily yields inequality $(1 - p_{v_i})(1 - p_{v_j}) \leqslant (1 - p_{v_k})(1 - p_{u_k})$ claimed. ∎

Lemma 5.7 has the following important consequence that will serve us to devise a polynomial-time algorithm for PROBABILISTIC MIN COLORING in graphs $\bar{M}_{n,n}$. Fix an optimal coloring C^* and consider the four leftmost vertices, say v_i, v_j and u_i, u_j of its two horizontal colors S_v and S_u, respectively (suppose that any of these colors has at most two vertices). Then, it suffices to determine vertical groups R_i and R_j (i.e., the ones where v_i, u_i and v_j, u_j, respectively, belong), so that the whole coloring C^* is determined. Indeed, applying Lemma 5.7 with v_i, v_j and u_i, u_j, the two leftmost vertices of colors S_v and S_u, respectively, the form of coloring C^* is as in Figure 5.9, i.e., colors on the left of v_i, u_i and from v_{i+1}, u_{i+1} to v_{j-1}, u_{j-1} are vertical, and the two horizontal colors have v_i, v_j and u_i, u_j as the leftmost vertices, the rest of them being $v_{j+1}, \ldots, v_n$ and $u_{j+1}, \ldots, u_n$, respectively.

The discussion above exhibits the following algorithm for PROBABILISTIC MIN COLORING in the class of graphs $\bar{M}_{n,n}$:

1) consider the full vertical coloring $(R_1, R_2, \ldots, R_n)$ and store its value;

2) consider the natural horizontal coloring (V, U) and store its value;

3) for $i = 1$ to n, consider coloring $(R_i, V \setminus R_i, U \setminus R_i)$ and store its value;

4) for $i = 1$ to n, for $j = i + 1$ to $n - 1$, consider the coloring of Figure 5.9 and store its value;

5) output the best of the colorings computed during Steps 1 to 4.

From what has been discussed previously, the coloring returned by Step 5 of the algorithm above is optimal for $\bar{M}_{n,n}$. Furthermore, it is easy to see that this algorithm runs in polynomial time (in fact, a careful implementation of it, including an ordering of vertical groups in decreasing order with respect to their weights $f(\cdot)$, implies a complexity of $O(n^2)$). Consequently the following theorem holds.

THEOREM 5.3.– PROBABILISTIC MIN COLORING *is polynomial for the bipartite complements of bipartite matchings.*

Another immediate consequence of Property 5.2 is the following proposition.

PROPOSITION 5.5.– *Assuming identical vertex-probabilities, the unique 2-coloring of* $\bar{M}_{n,n}$ *is optimal for* PROBABILISTIC MIN COLORING.

Proof. For PROBABILISTIC MIN COLORING, starting either from any mixed coloring, or from the vertical one, iterative application of Corollary 5.1 concludes that the unique 2-coloring is optimal. As we will see later, in section 5.7.1.2, this becomes not true when dealing with non-identical vertex-probabilities. ∎

5.4.4. *Trees*

Recall that the counter-example of Figure 5.5 shows that the natural 2-coloring is not always optimal in paths under distinct vertex-probabilities. In what follows, we study PROBABILISTIC MIN COLORING on trees. As previously, we assume that $|V| \geqslant |U|$.

PROPOSITION 5.6.– PROBABILISTIC MIN COLORING *can be optimally solved in trees with complexity bounded above by* $(n+1)^{\Delta(k\Delta+k+1)+1}$ *where* Δ *denotes the maximum degree of the tree and* k *the number of distinct vertex-probabilities.*

Proof. Consider a tree $T(N, E)$ of order n. Let $p_1, \ldots, p_k$ be the k distinct vertex-probabilities in T, n_i be the number of vertices of T with probability p_i and set $M = \prod_{i=1}^{k}\{0, \ldots, n_i\}$. Recall finally that, from Property 5.3, any optimal solution of PROBABILISTIC MIN COLORING in T uses at most $\Delta + 1$ colors.

Consider a vertex $v \in N$ with δ children and denote them by $v_1, \ldots, v_\delta$. Let $c \in \{1, \ldots, \Delta+1\}$ and $Q = \{q_1, \ldots, q_{\Delta+1}\} \in M^{\Delta+1}$ where, for any $j \in \{1, \ldots, \Delta+1\}$, $q_j = (q_{j_1}, \ldots, q_{j_k}) \in M$. We search to see if a coloring of $T[v]$ (i.e., of the subtree of T rooted at v) exists verifying both of the following properties:

– v is colored with color c;

– q_{i_j} vertices with probability p_i are colored with color j.

For this, let us define predicate $P_v(c, Q)$ with value *true* if such a coloring exists. In other words, we consider any possible configuration (in terms of number of vertices of any probability in any of the possible colors) for all the feasible colorings for $T[v]$.

One can determine value of P_v if one can determine values of P_{v_i}, $i = 1, \ldots, \delta$. Indeed, it suffices that one looks-up the several alternatives, distributing the q_{i_j} vertices (of probability p_i colored with color j) over the δ children of v (q_{i_j} may be $q_{i_j} - 1$ if $p(v) = p_i$ and $c = j$). More formally:

$$P_v(c, Q) = \bigvee_{(c_1,\ldots,c_\delta)} \bigvee_{(Q^1,\ldots,Q^\delta)} \left(P_{v_1}\left(c_1, Q^1\right) \wedge \ldots \wedge P_{v_\delta}\left(c_\delta, Q^\delta\right)\right) \qquad [5.31]$$

where in the clauses of [5.31]:

– for $j = 1, \ldots, \delta$, $c_j \neq c$ (in order that one legally colors v with color c),

– for $s = 1, \ldots, \delta$, $Q^s \in M^{\Delta+1}$, and

– for any pair (i, j):

$$\sum_{s=1}^{\delta} q_{i_j}^s = \begin{cases} q_{i_j} - 1 & \text{if } p(v) = p_i \text{ and } c = j \\ q_{i_j} & \text{otherwise} \end{cases}$$

Observe now that $|M| \leqslant (n+1)^k$ and, consequently, $|M^{\Delta+1}| \leqslant (n+1)^{k(\Delta+1)}$. For any vertex v, there exist at most $n|M^{\Delta+1}|$ values of P_v to be computed and for any of these computations, at most $(n|M^{\Delta+1}|)^\delta$ conjunctions, or disjunctions, have to be evaluated. Hence, the total complexity of this algorithm is bounded above by $n(n|M^{\Delta+1}|)^{\delta+1} \leqslant (n+1)^{\Delta(k\Delta+k+1)+1}$. To conclude, it suffices to output the coloring corresponding to the best of the values of predicate $P_r(c, Q)$, where r is the root of T. ∎

COROLLARY 5.8.– PROBABILISTIC MIN COLORING is polynomial in trees with bounded degree and with bounded number of distinct vertex-probabilities.

Since paths are trees of maximum degree 2, we get also the following result.

PROPOSITION 5.7.– PROBABILISTIC MIN COLORING *is polynomial in paths with bounded number of distinct vertex-probabilities. Consequently, it is polynomial for paths under identical vertex-probabilities.*

Let us note that for the second statement of Proposition 5.7, one can show something stronger, namely that *2-coloring is optimal for paths under identical vertex-probabilities*. Indeed, this case can be seen as an application of Property 5.6. The maximum independent set in a path coincides with V as any vertex of U is adjacent (and hence cannot have the same color) to a distinct vertex of V. This suffices to prove the proposition.

Consider now two particular classes of trees, denoted by $\mathcal{T}_E$ and $\mathcal{T}_O$, where all leaves lie exclusively either at even or at odd levels, respectively (root been considered at level 0). Trees in both classes can be polynomially checked. We are going to prove that, under identical vertex-probabilities, PROBABILISTIC MIN COLORING is polynomial for both $\mathcal{T}_E$ and $\mathcal{T}_O$. To do this, we first prove the following lemma where, for a tree T, we denote by N_E (resp., N_O) the even-level (resp., odd-level) vertices of T.

LEMMA 5.8.– *Consider $T \in \mathcal{T}_O$ (resp. in $\mathcal{T}_E$). Then N_O (resp., N_E) is a maximum independent set of T.*

Proof. We prove the lemma for $T \in \mathcal{T}_O$; case $T \in \mathcal{T}_E$ is completely similar. Set $n_o = |N_O|$, $n_e = |N_E|$ and note that $n_o > 0$ (otherwise, T consists of a single isolated vertex). We will show *ab absurdo* that there exists a maximum independent set S^* of T such that $S^* = N_O$ (resp., $S^* = N_E$).

Suppose *a contrario* that any independent set S^* satisfies $|S^*| > n_o$. Then the following two cases can occur:

$S^* \subseteq N_E$. This implies $|S^*| \leqslant n_e$. Since any vertex in N_E has at least a child, $n_e \leqslant n_o$, hence $|S^*| \leqslant n_o$, absurd since N_o is also an independent set and S^* is supposed to be the maximum one.

$S^* \subseteq N_O \cup N_E$. In other words, S^* contains vertices from both N_O and N_E. Then, for any vertex $e \in N_E \cap S^*$ that is parent of a leaf, e has at least a children with no other neighbors in S^*. We can then switch between S^* and its children, thus obtaining an independent set at least as large as S^*. We can iterate this argument with the vertices of this new independent set (denoted also by S^* for convenience) lying two levels above e (i.e., the great-grandparents of the leaves). Let g be such a vertex and assume that $g \in S^*$. Obviously, all its children are odd-level vertices and none of them is in S^* (*a contrario*, S^* would not be an independent set). Furthermore, none of these children can have a child $c \in S^*$ because e is an even-level vertex previously switched off from S^*, in order to be replaced by its children. Thus, we can again switch between g and its children, thus getting a new independent set S^* larger than the previous one. We again iterate up to the root, always obtaining a new "maximum independent set" larger than the older one. Moreover, at the end, the independent set obtained will verify $S^* = N_O$. ∎

PROPOSITION 5.8.– *Under identical vertex-probabilities,* PROBABILISTIC MIN COLORING *is polynomial in $\mathcal{T}_O$ and $\mathcal{T}_E$.*

Proof. By Lemma 5.8, trees in $\mathcal{T}_O$ and $\mathcal{T}_E$ fit Property 5.6. So, for these trees, 2-coloring is optimal. ∎ To conclude this paragraph, we deal with stars and show that PROBABILISTIC MIN COLORING is polynomial there under any probability system.

PROPOSITION 5.9.– *Under any vertex-probability system 2-coloring is optimal for stars.*

Proof. Note first that the center of the star constitutes a color *per se* in any feasible coloring. Then, Property 5.2 applied on star's leaves suffices to conclude the proof. ∎

5.4.5. *Cycles*

In what follows in this section, we deal with cycles C_n of size n with identical vertex-probabilities. We will prove that in such cycles, PROBABILISTIC MIN COLORING is polynomial. Let us note that, obviously, an odd cycle is not really a bipartite graph[2]. However, for economy, we integrate this case into this section also.

PROPOSITION 5.10.– PROBABILISTIC MIN COLORING *is polynomial in even cycles with identical vertex-probabilities.*

Proof. Note that in even cycles, Property 5.6 applies immediately; therefore, the natural 2-coloring is optimal. ∎

PROPOSITION 5.11.– PROBABILISTIC MIN COLORING *is polynomial in odd cycles with identical vertex-probabilities.*

Proof. Consider an odd cycle C_{2k+1}, denote by $1, 2, \ldots, 2k+1$ its vertices and fix an optimal solution C^* for it. By Property 5.3, $|C^*| \leqslant 3$. Since C_{2k+1} is not bipartite, we can immediately conclude that $|C^*| = 3$. Set $C^* = (S_1^*, S_2^*, S_3^*)$ and denote by S^* a maximum independent set of C_{2k+1}; assume $S^* = \{2i : i = 1, \ldots, k\}$, ie., $|S^*| = k$. By Property 5.2:

$$f(C^*) \geqslant f(S^*) + f_r^* = 1 - (1-p)^k + f_r \qquad [5.32]$$

where f_r^* is the value of the best coloring in the rest of C_{2k+1}, i.e., in the subgraph of C_{2k+1} induced by $V(C_{2k+1}) \setminus S^*$. This graph, of order $k+1$ consists of edge (v_1, v_{k+1}) and $k-1$ isolated vertices. Following, once more Property 5.2, in a graph of order $k+1$ that is not a simple set of isolated vertices, the ideal coloring would be an independent set of size k and a singleton of total value $1-(1-p)^k+p$. So, using [5.32], we get: $f(C^*) \geqslant 2-2(1-p)^k+p$. But the coloring $\hat{C} = (S^*, \{2i-1 : i = 1, \ldots, k\}, \{2k+1\})$ attains this value; therefore, it is optimal for C_{2k+1}, QED. ∎

2. Recall that a graph is bipartite, if and only if, it does not have odd cycles (see section A.2 in Appendix A).

5.5. Complements of bipartite graphs

Given a bipartite graph $B(V, U, E)$, its complement, $\bar{B}(V, U, \bar{E})$, is a loopless graph consisting of two cliques, one on V and one on U, plus the set of edges $\bar{E}' = \{v_i u_j \notin E : v_i \in V, u_j \in U\}$; in other words, the edges of $\bar{E}$ are the edges of the cliques $K_{|V|}$ and $K_{|U|}$ and the edges between V and U missing from E. The graph in Figure 5.10(b) is the complement of the bipartite graph of Figure 5.10(a). These graphs have the property that any independent set is of cardinality at most 2. In other words, any coloring there is a collection of independent sets of size 2 and of singletons. The following lemma characterizes the functional's value of a such a coloring.

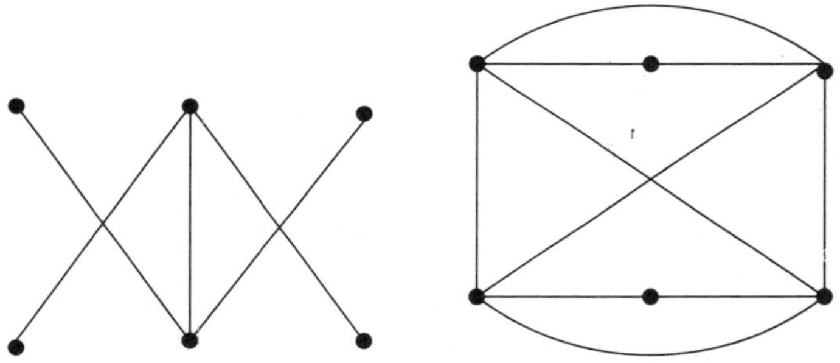

(a) B (b) $\bar{B}$

Figure 5.10. *A bipartite graph B (Figure 5.10(a)) and its complement $\bar{B}$ (Figure 5.10(b))*

LEMMA 5.9.– *Let $\bar{B}$ be the complement of bipartite graph B, let n be the order of $\bar{B}$ and B, let C be a coloring of $\bar{B}$ and let $S = \{\{v_{i_k}, u_{j_k}\} : k = 1, \ldots, |S|\}$ be the collection of independent sets of size 2 in C. Then $E(\bar{B}, C) = \sum_{i=1}^{n} p_i - \sum_{k=1}^{|S|} p_{i_k} p_{j_k}$.*

Proof. The value of a color $\{v_{i_k}, u_{j_k}\} \in C$ will be $1 - (1 - p_{i_k})(1 - p_{j_k}) = p_{i_k} + p_{j_k} - p_{i_k} p_{j_k}$; on the other hand, the value of a singleton $\{v_i\} \in C$ will be p_i. Consequently, it is easy to see that the functional of C will be $E(\bar{B}, C) = \sum_{i=1}^{n} p_i - \sum_{k=1}^{|S|} p_{i_k} p_{j_k}$ as claimed. ∎

From Lemma 5.9, the first term of $E(\bar{B}, C)$ is constant; so, $E(\bar{B}, C)$ is minimized when its second term is maximized. Consider the bipartite graph $B'(V, U, E(B'))$ with $E(B') = (V \times U) \setminus \bar{E}'$ and assign to any edge $v_i u_j \in E(B')$ weight $p_i p_j$. Then, collection S becomes a matching of B' and the term $\sum_{k=1}^{|S|} p_{i_k} p_{j_k}$ is the total weight of this matching. Recall finally that a maximum weight matching can be polynomially computed in any graph ([PAP 81]). Then, consider the following algorithm: given $\bar{B}$:

– transform it into B' and weight any of its edges with the product of the probabilities of its endpoints;

– compute a maximum weight matching M in B';

– color endpoints of any edge of M with an unused color (the same for both endpoints);

– color the remaining vertices of B' with an unused color by such vertex.

From what has been discussed, the coloring so produced is optimal and the following result holds immediately.

THEOREM 5.4.– PROBABILISTIC MIN COLORING *is polynomial in complements of bipartite graphs.*

5.6. Split graphs

5.6.1. *The complexity of* PROBABILISTIC MIN COLORING

We deal now with split graphs. This class of graphs is quite close to bipartite ones, since any split graph of order $m + n$ is composed by a clique K_m, on m vertices, an independent set S of size n and some edges linking vertices of $V(K_m)$ to vertices of S. In Figure 5.11, such a graph is illustrated for $m = 3$ and $n = 4$.

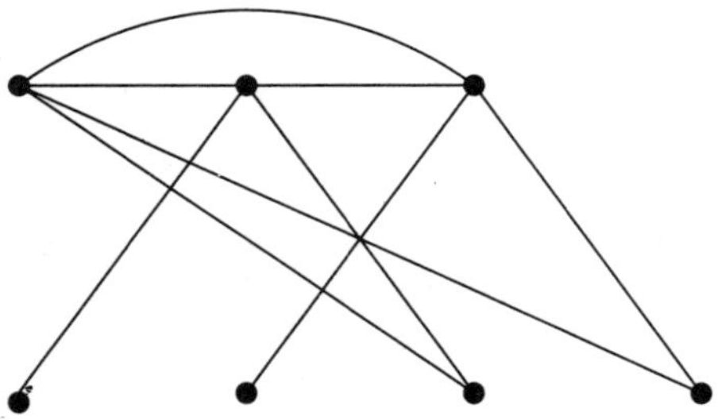

Figure 5.11. *A split graph with $m = 3$ and $n = 4$*

These graphs are, in some senses, midway between bipartite graphs and complements of bipartite graphs. In what follows, we first show that PROBABILISTIC MIN COLORING is **NP**-hard in split graphs even under identical vertex-probabilities. For this, we prove that the decision counterpart of PROBABILISTIC MIN COLORING in split graphs is **NP**-complete. This counterpart, denoted by PROBABILISTIC MIN COLORING(K), is defined as follows: "given a split graph $G(V, E)$, a system of identical vertex-probabilities for G and a constant $K \leqslant |V|$, does there exist a coloring the functional of which is at most K?".

THEOREM 5.5.– PROBABILISTIC MIN COLORING(K) *is **NP**-complete in split graphs, even assuming identical vertex-probabilities.*

Proof. Inclusion of PROBABILISTIC MIN COLORING(K) in **NP** is immediate. In order to prove completeness, we will reduce 3-EXACT COVER ([GAR 79]) to our problem. Given a family $\mathcal{S} = \{S_1, S_2, \ldots, S_m\}$ of subsets of a ground set $\Gamma = \{\gamma_1, \gamma_2, \ldots, \gamma_n\}$ (we assume that $\cup_{S_i \in \mathcal{S}} S_i = \Gamma$) such that $|S_i| = 3$, $i = 1, \ldots, m$, we are asked if there exists a subfamily $\mathcal{S}' \subseteq \mathcal{S}$, $|\mathcal{S}'| = n/3$, such that $\mathcal{S}'$ is a partition on Γ. Obviously, we assume that n is a multiple of 3.

Consider an instance $(\mathcal{S}, \Gamma)$ of 3-EXACT COVER and set $q = n/3$. The split graph $G(V, E)$ for PROBABILISTIC MIN COLORING will be constructed as follows:

– family $\mathcal{S}$ is replaced by a clique K_m (i.e., we take a vertex per set of $\mathcal{S}$); denote by $s_1, \ldots, s_m$ its vertices;

– ground set Γ is replaced by an independent set $X = \{v_1, \ldots, v_n\}$;

– $(s_i, v_j) \in E$ iff $\gamma_j \notin S_i$;

– $p > 1 - (1/q)$;

– $K = mp + q(1-p) - q(1-p)^4$.

Figure 5.12 illustrates the split graph obtained, by application of the three first items of the construction above, on the following 3-EXACT COVER-instance:

$$
\begin{aligned}
\Gamma &= \{\gamma_1, \gamma_2, \gamma_3, \gamma_4, \gamma_5, \gamma_6\} \\
\mathcal{S} &= \{S_1, S_2, S_3, S_4, S_5\} \\
S_1 &= \{\gamma_1, \gamma_2, \gamma_3\} \\
S_2 &= \{\gamma_1, \gamma_2, \gamma_4\} \\
S_3 &= \{\gamma_3, \gamma_4, \gamma_5\} \\
S_4 &= \{\gamma_4, \gamma_5, \gamma_6\} \\
S_5 &= \{\gamma_3, \gamma_5, \gamma_6\}
\end{aligned}
\qquad [5.33]
$$

Suppose that a partition $\mathcal{S}' \subseteq \mathcal{S}$, $|\mathcal{S}'| = q = n/3$ is given for $(\mathcal{S}, \Gamma, q)$. Order $\mathcal{S}$ in such a way that the q first sets are in $\mathcal{S}'$. For any $S_i \in \mathcal{S}'$, set $S_i = \{\gamma_{i_1}, \gamma_{i_2}, \gamma_{i_3}\}$. Then, subset $\{s_i, v_{i_1}, v_{i_2}, v_{i_3}\}$ of V is an independent set of G. Construct for G the coloring $C = (\{s_i, v_{i_1}, v_{i_2}, v_{i_3}\}_{i=1,\ldots,q}, \{s_{q+1}\}, \ldots, \{s_m\})$. It is easy to see that $f(C) = q(1-(1-p)^4) + (m-q)p = mp + q(1-p) - q(1-p)^4 = K$.

Conversely, suppose that a coloring C is given for G with value $f(C) \leqslant K$. There exist, in fact, two types of feasible coloring in G:

1) C is as described just above, i.e., it is of the form:

$$C = \left(\{s_i, v_{i_1}, v_{i_2}, v_{i_3}\}_{i=1,\ldots,q}, \{s_{q+1}\}, \ldots, \{s_m\}\right)$$

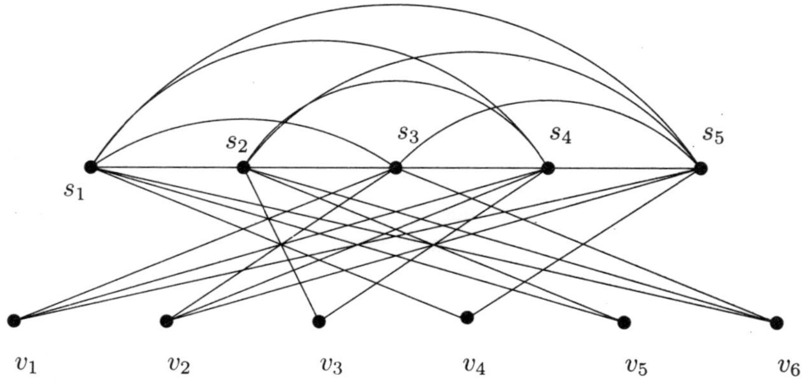

Figure 5.12. *The split graph obtained from* 3-EXACT COVER-*instance described in [5.33]*

2) up to reordering of colors, C is of the form:

$$C = (S_1, \ldots, S_{q_4}, S_{q_4+1}, \ldots, S_{q_4+q_3}, S_{q_4+q_3+1}, \ldots, S_{q_4+q_3+q_2}, \{v_{q_4+q_3+q_2+1}\}, \ldots, \{v_m\}, X') \quad [5.34]$$

where:

- the q_4 first sets are of the form: $\{s_i, v_{i_1}, v_{i_2}, v_{i_3}\}$, $i = 1, \ldots, q_4$;
- the q_3 next sets are of the form: $\{s_i, v_{i_1}, v_{i_2}\}$, $i = q_4 + 1, \ldots, q_4 + q_3$;
- the q_2 next sets are of the form: $\{s_i, v_{i_1}\}$, $i = q_4 + q_3 + 1, \ldots, q_4 + q_3 + q_2$:
- the $m - (q_4 + q_3 + q_2)$ singletons are the remaining vertices of K_m which form a color per such vertex;
- X' is the subset of X not contained in the colors above.

We note that coloring $C' = (\{s_1\}, \ldots, \{s_m\}, X)$ is a particular case of [5.34] with $q_1 = q_2 = q_3 = 0$.

If C is of Type 1, then for any color $\{s_i, v_{i_1}, v_{i_2}, v_{i_3}\}$, $i = 1, \ldots, q$, we take set S_i in $\mathcal{S}'$. By construction of G, set S_i covers elements γ_{i_1}, γ_{i_2} and γ_{i_3} of the ground set Γ. The q sets thus selected form a partition on Γ of cardinality q.

Let us now assume that C is of Type 2 (see [5.34]). Note first that, for coloring C' mentioned at the end of Item 2 above, and for $p > 1 - (1/q)$:

$$f(C') = mp + (1 - (1-p)^n) > mp + q(1-p) - q(1-p)^4 = K \quad [5.35]$$

Note first that color X' (see Item 2) can never satisfy $|X'| \geqslant 4$; *a contrario*, using the local optimality argument of Property 5.4, since X' is the largest color, coloring C' would have value smaller than the one of C; hence the latter value would be greater than K (see [5.35]). Therefore, we can assume $|X'| \leqslant 3$. In this case, one can, by keeping the q_4 colors of size 4 unchanged, progressively transform the rest of the colors by successive applications of Property 5.4 in order to create new (possibly unfeasible) 4-colors. This can be done by moving vertices from the smaller colors to the larger ones and is always possible since $n - 3q_4$ is a multiple of 3. Therefore, at the end of this processus, one can obtain exactly q (possibly unfeasible) 4-colors, the remaining vertices being colored with one color by vertex. Denoting by C'' the "coloring" thus obtained, we have obviously, $f(C'') = K < f(C)$.

Therefore, from the discussion above, the only coloring having value at most K is the one of Type 1, QED. ∎

Split graphs are particular cases of larger graph-family, the chordal graphs (graphs for which any cycle of length at least 4 has a chord ([BER 73])).

COROLLARY 5.9.– PROBABILISTIC MIN COLORING is **NP**-hard in chordal graphs even under identical vertex-probabilities.

5.6.2. *Approximation results*

We deal in this section with the approximability of PROBABILISTIC MIN COLORING in split graphs. Let $G(K, S, E)$ be such a graph, where K is the vertex set of the clique ($|K| = m$) and S is the independent set ($|S| = n$). Fix an optimal PROBABILISTIC MIN COLORING-solution $C^* = (S_1^*, S_2^*, \dots, S_{k^*}^*)$ in $G(K, S, E)$.

LEMMA 5.10.– $m \leqslant k^* \leqslant m + 1$.

Proof. Since vertex-set K forms a clique, any solution in G will use at least m colors. On the other hand, if C^* uses more than m colors, this is due to the fact that there exist elements of S that cannot be included in any of the m colors associated with the vertices of K. If at least two such colors are used, then, since both of them are proper subsets of S (recall that S is an independent set), the local optimality argument of Property 5.1 would conclude the existence of a solution better than C^*, which is a contradiction. ∎

Consider first the natural coloring, denoted by C, consisting of taking an unused color for any vertex of K and a color for the whole set S (in other words C uses $m+1$ colors for G).

PROPOSITION 5.12.– *Coloring C is a 2-approximation for split graphs under any system of vertex-probabilities.*

Proof. Denote by $C^* = (S_1^*, S_2^*, \ldots, S_{k^*}^*)$, an optimal solution in G and assume that colors are ranged in decreasing-value order, i.e., $f(S_i^*) \geqslant f(S_{i+1}^*)$, $i = 1, \ldots, k^* - 1$. From Lemma 5.10, $m \leqslant k^* \leqslant m + 1$. If $k^* = m + 1$ and S_1^* is the color that is a subset of S, then local optimality arguments of Property 5.2 conclude that C is optimal. Hence, assume that S_1^* is a color including a vertex of K and vertices of S. For reasons of facility assume also that, upon a reordering of vertices, vertex $v_i \in K$ is included in color S_i^*; also, denote by p_i, the probability of vertex $v_i \in K$ and by q_i the probability of a vertex $v_i \in S$. Then:

$$f(C) = \sum_{i=1}^{m} p_i + \left(1 - \prod_{i=1}^{n} (1 - q_i)\right) \quad [5.36]$$

$$f(C^*) \geqslant \sum_{i=2}^{m} p_i + \left(1 - (1 - p_1) \prod_{i=1}^{n} (1 - q_i)\right) \quad [5.37]$$

where [5.37] holds thanks to local optimality arguments leading to Property 5.2, when we charge color S_1^* with all vertices of S. Observe also that:

$$1 - \prod_{i=1}^{n} (1 - q_i) \leqslant 1 - (1 - p_1) \prod_{i=1}^{n} (1 - q_i) \quad [5.38]$$

$$1 - (1 - p_1) \prod_{i=1}^{n} (1 - q_i) \geqslant p_1 \quad [5.39]$$

Combining [5.36] and [5.37], and using also [5.38] and [5.39], we get:

$$\frac{f(C)}{f(C^*)} \leqslant \frac{p_1 + \sum_{i=2}^{m} p_i + \left(1 - \prod_{i=1}^{n} (1 - q_i)\right)}{\sum_{i=2}^{m} p_i + \left(1 - (1 - p_1) \prod_{i=1}^{n} (1 - q_i)\right)}$$

$$\overset{[5.38]}{\leqslant} \frac{p_1 + \sum_{i=2}^{m} p_i + \left(1 - (1 - p_1) \prod_{i=1}^{n} (1 - q_i)\right)}{\sum_{i=2}^{m} p_i + \left(1 - (1 - p_1) \prod_{i=1}^{n} (1 - q_i)\right)}$$

$$= 1 + \frac{p_1}{\sum_{i=2}^{m} p_i + \left(1 - (1 - p_1) \prod_{i=1}^{n} (1 - q_i)\right)}$$

$$\overset{[5.39]}{\leqslant} 1 + \frac{p_1}{p_1 + \sum_{i=2}^{m} p_i} \leqslant 2$$

and the proof of the proposition is complete. ∎

We now show the main positive approximation result of this section, namely that PROBABILISTIC MIN COLORING in split graphs can be solved by a polynomial time approximation schema, under any system of vertex-probabilities.

THEOREM 5.6.– PROBABILISTIC MIN COLORING *in split graphs is approximable by a polynomial time approximation schema.*

Proof. Consider a split graph $G(K, S, E)$ and some optimal coloring $C^* = (S_1^*, S_2^*, \ldots)$ of G, with $f(S_1^*) \geqslant f(S_2^*) \geqslant \ldots$. Assume, without loss of generality, that C^* contains:

– some colors built from one vertex of K and some vertices of S;

– some singletons of vertices of K;

– less than one color all of its vertices belong to S (by Corollary 5.1); we denote this color by S_r^*.

Then the following facts can be derived for the form of C^*:

1) for any $i > r$, S_i^* is a singleton $\{k_{j_i}\} \subset K$;

2) for every $i < r$, the independent set (color) S_i^* is maximal (for the inclusion) for the graph $G_i = G[V \setminus (S_1^* \cup \ldots \cup S_{i-1}^*)]$ (where, for $i = 1$, $S_{i-1}^* = \emptyset$).

Indeed, for Fact 1, if there exists a color $S_i^* = \{k_{j_i}, s_i^1, s_i^2, \ldots\}$, for $i > r$, then by the local optimality arguments of Corollary 5.1 and given the ordering assumed for the colors $S_1^*, S_2^*, \ldots$, putting vertices $s_i^1, s_i^2, \ldots$ in S_r^* would improve the value of C^*.

On the other hand, for Fact 2, if S_i^* is not maximal for G_i, there exists a color S_j^*, $j > i$, some vertices of which can be legally introduced in S_i^*. Given the ordering of the colors, introduction of these vertices in S_i^* would lead to improvement of the value of C^*. Note now that, one can conclude from Fact 2 that, if S_i^* is not a singleton k_i of K, but it also contains some vertices of $S \setminus (S_1^* \cup \ldots \cup S_{i-1}^*)$, then it contains all the vertices of $S \setminus (S_1^* \cup \ldots \cup S_{i-1}^*)$ that are not its neighbors. This implies that if one could know exactly which vertex of K belongs to color S_i^*, then one can exactly determine any color S_i^*, for $i < r$.

Now, given a subset X of k distinct vertices of K, we denote by C_X the set of the k colors built following the rule of Fact 2. Consider also coloring C consisting of taking an unused color for any vertex of K and a color for the whole set S (i.e., the one studied in Proposition 5.12). Revisit the proof of Proposition 5.12 and note that from [5.37], [5.38] and [5.39]:

$$f\left(C^*\right) \geqslant f(C) - f\left(S_1^*\right) \qquad [5.40]$$

Consider the following algorithm SCHEMA (it is rather a family of algorithms parameterized by a constant $\epsilon > 0$):

1) fix an $\epsilon > 0$;

2) set $k = \lceil 1/\epsilon \rceil$;

3) build and store coloring C of Proposition 5.12 for G;

4) for any $k' \in \{1, \ldots, k-1\}$ and any set $X \subset K$, such that $|X| = k'$:

a) construct the k'-coloring C_1 derived by the vertices of X along the rules of Fact 2;

b) consider the subgraph of G induced by the still uncolored vertices and built the coloring C_2 of Proposition 5.12 for this graph;

c) build and store coloring $C' = (C_1, C_2)$;

5) output the best coloring $\hat{C}$ among coloring C and the colorings C' built in Steps 3 and 4c, respectively.

All executions of Step 4 need at most $O(m^k) = O(m^{\lceil 1/\epsilon \rceil})$ while an execution of Steps 4a) and 4b) take at most $O(nm)$. So, the overall complexity of SCHEMA is in $O(nm^{1+(1/\epsilon)})$, polynomial if ϵ is fixed.

Note first that if $r \leqslant k$, then $\hat{C}$ is optimal. Indeed, any subset of K of size $r-1$ has been processed during the iterations of Step 4 and any of the colorings C_X obtained has been completed by the still uncolored part of S (constituting a color) and by as many colors as the yet uncolored vertices of K. By what has been discussed above, in Facts 1 and 2 and just after them, one of the colorings so-built and completed is optimal and has been retained by SCHEMA.

If, on the other hand, $r > k$, then for the set X^*, corresponding to C^*, the coloring C_{X^*} obtained is $C_{X^*} = \{S_1^*, S_2^*, \ldots, S_{k-1}^*\}$. Furthermore, on the subgraph of G induced by the still uncolored vertices, the coloring C_2 obtained is such that (consider [5.40]):

$$f(C_2) \leqslant f(S_k^*, \ldots, S_\ell^*) + f(S_k^*) \qquad [5.41]$$

where ℓ denotes the number of colors in C^*. Using [5.41], we get:

$$f\left(\hat{C}\right) \leqslant f(C^*) + f(S_k^*) \leqslant f(C^*)\left(1 + \frac{1}{k}\right) \leqslant f(C^*)(1+\epsilon)$$

In other words, for any $\epsilon > 0$:

$$\frac{f\left(\hat{C}\right)}{f(C^*)} \leqslant 1 + \epsilon$$

So, for a fixed $\epsilon > 0$, SCHEMA constitutes a polynomial time approximation schema for PROBABILISTIC MIN COLORING in split graphs. ∎

Now go back to the proof of Theorem 5.5 and notice that it works for any $p > 1-(1/q)$, where $q = n/3$. Denote by $|G|$, the size of G in a suitable encoding. Notice finally that, given that $|X'| \leqslant 3$, application of the local optimality principle of Property 5.4, in the case where the initial instance of 3-EXACT COVER is a *yes*-instance (see [GAR 79]), the second best solution, for G is coloring $C' = (\{s_1\}, \ldots, \{s_m\}, X)$ with value $f(C') = mp+1-(1-p)^n$; furthermore, C' is feasible in any split graph.

Assume that a fully polynomial time approximation schema $\mathtt{A}_\epsilon$ exists for PROBABILISTIC MIN COLORING in split graphs. Consider a graph G, resulting from the transformation described in the proof of Theorem 5.5 from an instance $(\mathcal{S}, \Gamma)$ of 3-EXACT COVER, with $p > 1-(3/n)$ say $p = 1-(1/\omega(n))$, where ω is some polynomial with positive coefficients. Apply $\mathtt{A}_\epsilon$ to G and take as the final solution the best among the solutions computed by this schema and C'.

If $(\mathcal{S}, \Gamma)$ is a *no*-instance, then C' is an optimal solution for G.

Suppose now that $(\mathcal{S}, \Gamma)$ is a *yes*-instance. In this case, the best coloring for G has value K and C' achieves ratio:

$$\begin{aligned}
\frac{f(C')}{K} &= \frac{mp+1-(1-p)^n}{mp+q(1-p)-q(1-p)^4} \\
&= \frac{mp+1-(1-p)^n}{mp+q(1-p)\left(1-(1-p)^3\right)} \\
&\geqslant \frac{mp+1-(1-p)^4}{mp+1-(1-p)^3} = 1+\frac{p(1-p)^3}{mp+1-(1-p)^3}
\end{aligned}$$

Henceforth, execution of $\mathtt{A}_\epsilon$ on G with $\epsilon < p(1-p)^3/(mp+1-(1-p)^3)$ will return the optimal coloring of G with value K and, in this case, one can safely answer that $(\mathcal{S}, \Gamma)$ is a *yes*-instance for 3-EXACT COVER. Notice finally that since $p = 1-(1/\omega(n))$, $\epsilon \approx 1/m\omega^3(n)$, i.e., $1/\epsilon \approx m\omega^3(n)$. So, $\mathtt{A}_\epsilon$ becomes an optimal and polynomial algorithm correctly deciding 3-EXACT COVER.

As a consequence, the following concluding proposition can be immediately derived from the discussion above.

THEOREM 5.7.– *Unless* ***P*** $=$ ***NP***, PROBABILISTIC MIN COLORING *on split graphs cannot be solved by a fully polynomial time approximation schema.*

5.7. Determining the best k-coloring in k-colorable graphs

In what follows we focus on an interesting variant of coloring, the one where we fix an integer k and wish to determine the best k-coloring in a k-colorable graph. In particular, we tackle this problem in bipartite graphs and in their complements.

5.7.1. *Bipartite graphs*

We deal here with the most popular k-colorable graphs, for any $k \geqslant 2$, the bipartite graphs. As previously, we denote by $B(V, U, E)$ a connected bipartite graph with bipartition V and U and edge-set E. The problem covered in this section, denoted by PROBABILISTIC MIN k-COLORING in what follows, consists, for a k greater than 2 and smaller than the order n of the input bipartite graph B, of determining the best among the k-colorings of B (since B is bicolorable, there exists at least one coloring for any such value of k; on the other hand, an n-coloring of any graph is trivial). We show that PROBABILISTIC MIN k-COLORING is **NP**-hard for any $k \geqslant 3$.

5.7.1.1. PROBABILISTIC MIN 3-COLORING

In order to prove the **NP**-hardness of PROBABILISTIC MIN k-COLORING, we first need to prove an initial completeness result about PROBABILISTIC MIN 3-COLORING that will serve us as a basis. Obviously, PROBABILISTIC MIN 3-COLORING consists of determining the best among the 3-colorings of a bipartite graph B. We denote by PROBABILISTIC MIN 3-COLORING(K) the decision version of PROBABILISTIC MIN 3-COLORING, i.e., the one where for some constant K, we search for determining if there exists a 3-coloring of B with a value at most K. We first prove that PROBABILISTIC MIN 3-COLORING(K) is **NP**-complete. Then, we extend the gadget that we use for this proof and we use properties of it in order to form a recursive argument helping us to show that PROBABILISTIC MIN k-COLORING is **NP**-hard for any k.

PROPOSITION 5.13.– PROBABILISTIC MIN 3-COLORING(K) *is **NP**-complete.*

Proof. PROBABILISTIC MIN 3-COLORING(K) is obviously in **NP**. The completeness will be proved by reduction from the following precoloring extension problem, called 1-PREXT (shown to be **NP**-complete in [BOD 94]): *"given a bipartite graph $B(V, U, E)$ with $|V \cup U| \geqslant 3$ and three vertices v_1, v_2, v_3, does there exist a 3-coloring (S_1, S_2, S_3) of B such that $v_i \in S_i$ for $i = 1, 2, 3$?"*.

Consider an instance $B'(V, U', E', v_1, v_2, v_3)$ of 1-PREXT and note that we can assume that v_1, v_2, v_3 all belong either to V or to U'; in the opposite case it is easy to see that 1-PREXT is polynomial. Suppose that v_1, v_2, v_3 are in V. We transform $B'(V, U', E', v_1, v_2, v_3)$ into an instance of PROBABILISTIC MIN 3-COLORING(K) in the following way:

– add in U' three new vertices u_1, u_2, u_3 and set $U = U' \cup \{u_1, u_2, u_3\}$; add in E' the edge-set $E'' = \{v_i u_j : i, j = 1, 2, 3, i \neq j\}$ and set: $E = E' \cup E''$ and $B = B(V, U, E)$;

– the probability vector $\mathbf{Pr}$ is as follows: $p(u_1) = p(v_1) = \epsilon$, $p(u_2) = p(v_2) = \epsilon^2$, $p(u_3) = p(v_3) = \epsilon^3$, for $\epsilon \leqslant 1/10$, $p(v_i) = 0$, $v_i \in (V \cup U) \setminus \{v_i, u_i : i = 1, 2, 3\}$;

– set $K = 2\epsilon + \epsilon^2 + 2\epsilon^3 - \epsilon^4 - \epsilon^6$.

Obviously, the transformation of B' into B can be performed in polynomial time. We claim that $(B, \mathbf{Pr}, 3, K)$ has a 3-coloring with functional at most $2\epsilon + \epsilon^2 + 2\epsilon^3 - \epsilon^4 - \epsilon^6$ if and only if we can 3-color $B'(V, U', E', v_1, v_2, v_3)$ by assigning any of v_1, v_2, v_3 with a distinct color.

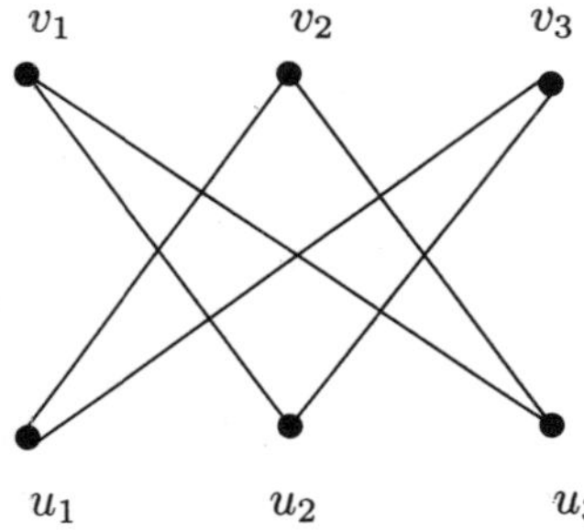

Figure 5.13. *The graph* $\bar{M}_{3,3} = B[\{v_i, u_i : i = 1, 2, 3\}]$

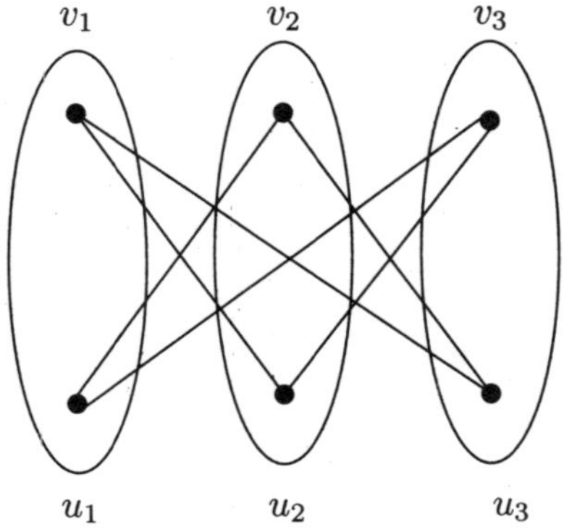

Figure 5.14. *The 3-coloring* C^* *restricted to* $\bar{M}_{3,3}$;
$E(B, C^*) = 2\epsilon + \epsilon^2 + 2\epsilon^3 - \epsilon^4 - \epsilon^6$

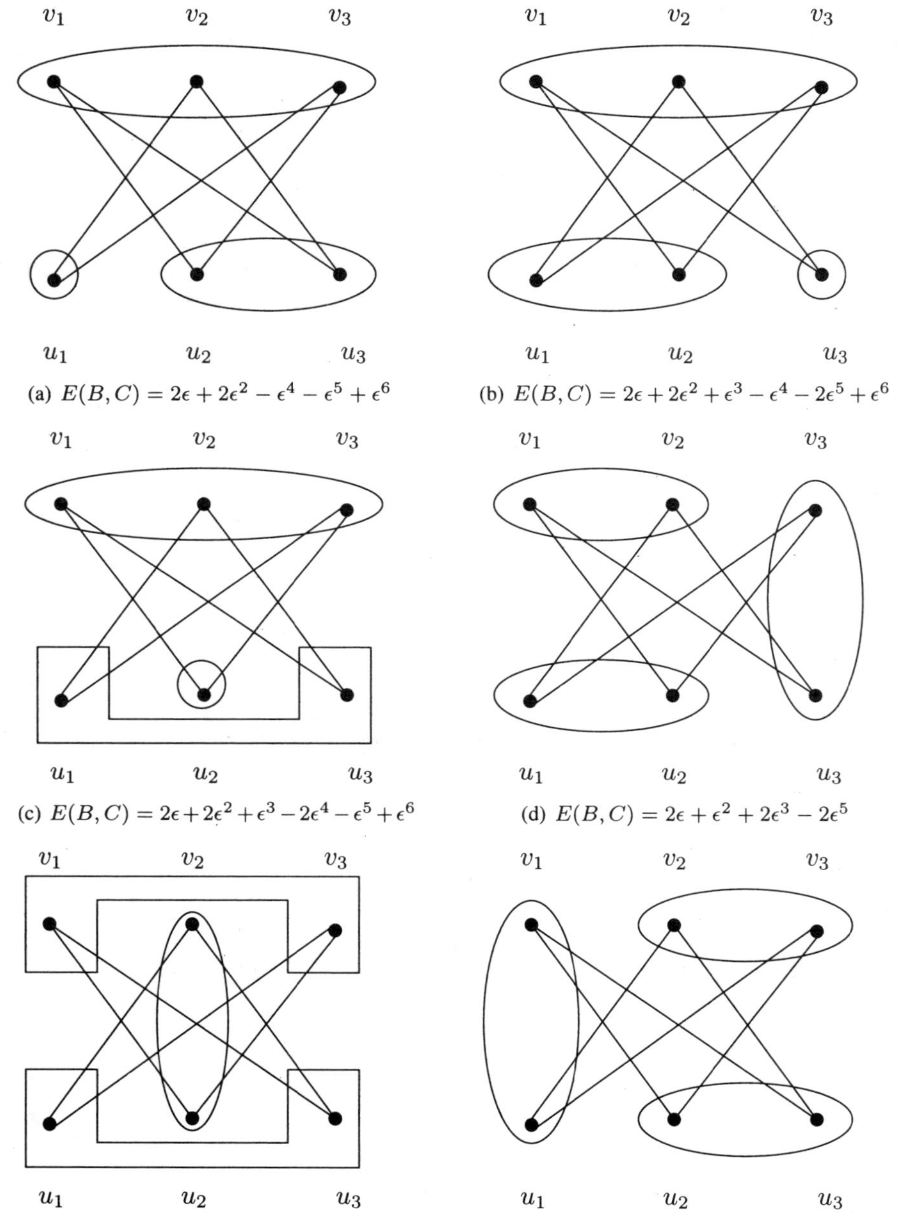

(a) $E(B,C) = 2\epsilon + 2\epsilon^2 - \epsilon^4 - \epsilon^5 + \epsilon^6$

(b) $E(B,C) = 2\epsilon + 2\epsilon^2 + \epsilon^3 - \epsilon^4 - 2\epsilon^5 + \epsilon^6$

(c) $E(B,C) = 2\epsilon + 2\epsilon^2 + \epsilon^3 - 2\epsilon^4 - \epsilon^5 + \epsilon^6$

(d) $E(B,C) = 2\epsilon + \epsilon^2 + 2\epsilon^3 - 2\epsilon^5$

(e) $E(B,C) = 2\epsilon + 2\epsilon^2 + 2\epsilon^3 - 3\epsilon^4$

(f) $E(B,C) = 2\epsilon + 2\epsilon^2 - \epsilon^6$

Figure 5.15. *The 6 distinct-value non-optimal 3-colorings of $B[\{v_i, u_i : i = 1, 2, 3\}]$ with the values of the functionals associated*

It is easy to see that the contribution of any vertex in $(V \cup U) \setminus \{v_i, u_i : i = 1, 2, 3\}$ in any coloring of B is null. Denote by $\bar{M}_{3,3}$ the graph $B[\{v_i, u_i : i = 1, 2, 3\}]$ (this graph is a kind of bipartite complement of a perfect matching on 3 edges, in our case on edges $v_i u_i$, $i = 1, 2, 3$) and observe that the (non-zero) value of any coloring of B is value of some coloring of $\bar{M}_{3,3}$. In Figure 5.13, the graph $\bar{M}_{3,3}$ is shown. Observe also that the value of K, introduced in the third item above, corresponds to a 3-coloring C^* of B taking $\{v_i, u_i\}$, $i = 1, 2, 3$ in the same color, say S_i; this coloring has functional equal to $\sum_{i=1}^{3}(1-(1-\epsilon^i)^2) = 2\epsilon+\epsilon^2+2\epsilon^3-\epsilon^4-\epsilon^6$. In Figure 5.14, the optimal functional-value 3-coloring C^* restricted to $\bar{M}_{3,3}$ is presented. By a simple inspection of any other 3-coloring C of B (there exist seven distinct functional-value colorings), one can easily see that the functional of C is, for $\epsilon \leqslant 1/10$, greater than the functional K of C^*. In Figure 5.15, the other non-optimal 3-colorings of $\bar{M}_{3,3}$ are illustrated.

So, if a 3-coloring C^* of B is polynomially computed with functional $K = 2\epsilon + \epsilon^2 + 2\epsilon^3 - \epsilon^4 - \epsilon^6$, then C^* restricted to $\bar{M}_{3,3}$ is in the form of Figure 5.14 (recall that the contribution of the vertices of $(V \cup U) \setminus \{v_i, u_i : i = 1, 2, 3\}$ in any coloring of B is 0). Consequently, C^* 3-colors the vertices of B' by assigning a distinct color to each of v_1, v_2, v_3.

Conversely, if a 3-coloring assigning a distinct color, say S_1, S_2 and S_3 to each of v_1, v_2, v_3, respectively, is computed for B', then $(S_1 \cup \{u_1\}, S_2 \cup \{u_2\}, S_3 \cup \{u_3\})$ is a coloring for B with functional $K = 2\epsilon + \epsilon^2 + 2\epsilon^3 - \epsilon^4 - \epsilon^6$. ∎

REMARK 5.2.– The functional of the (unique) 2-coloring of $B[\{v_i, u_i : i = 1, 2, 3\}]$ (figure 5.16) has value $2(\epsilon+\epsilon^2-\epsilon^4-\epsilon^5+\epsilon^6) > 2\epsilon+\epsilon^2+2\epsilon^3-\epsilon^4-\epsilon^6$, for $\epsilon \leqslant 1/10$.

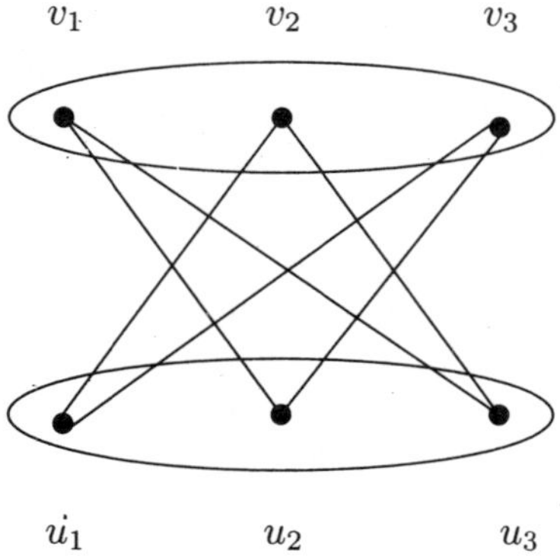

Figure 5.16. *The 2-coloring of* $B[\{v_i, u_i : i = 1, 2, 3\}]$ *with functional of value* $2(\epsilon + \epsilon^2 - \epsilon^4 - \epsilon^5 + \epsilon^6)$

So, the following theorem concludes the discussion of this section.

THEOREM 5.8.– PROBABILISTIC MIN 3-COLORING *is **NP**-hard.*

5.7.1.2. PROBABILISTIC MIN k-COLORING *for* $k > 3$

We now consider PROBABILISTIC MIN k-COLORING, where we look for the best k-coloring for any $k \in \{4, \ldots, n\}$ and its decision version PROBABILISTIC MIN k-COLORING(K). We will establish that PROBABILISTIC MIN k-COLORING(K) is **NP**-complete for any such k.

Consider a bipartite complement of a perfect matching with k edges[3].

The **NP**-hardness of PROBABILISTIC MIN k-COLORING is based upon the following proposition.

PROPOSITION 5.14.– *Consider $\bar{M}_{k,k}$ and assume that there exists a vertex-probability system* $\mathbf{Pr}$ *with $p(v_i) = p(u_i) = p_i$, $p_i \neq p_j$, $i, j = 1, \ldots, k$, such that, for any i, $3 \leqslant i < k$, the functional of a vertical coloring of any subgraph $\bar{M}_{i,i}$ of $\bar{M}_{k,k}$ is (strictly) smaller than the functional of the 2-coloring of $\bar{M}_{i,i}$. Consider a k-coloring C of $\bar{M}_{k,k}$ with value equal to $\sum_{i=1}^{k}(1-(1-p_i)^2)$. Then this coloring is vertical, i.e., of the form $\{\{v_i, u_i\} : i = 1, \ldots, k\}$ and the functional associated with it is the smallest over the functional of any feasible coloring of $\bar{M}_{k,k}$.*

Proof. Set $C = (S_1, \ldots, S_k)$, and remove vertices v_k and u_k from $\bar{M}_{k,k}$ together with their incident edges, in order to obtain graph $\bar{M}_{k-1,k-1}$. Then, the following three cases can appear: **(a)** $\bar{M}_{k-1,k-1}$ remains colored with k colors; **(b)** $\bar{M}_{k-1,k-1}$ is colored with $k-1$ colors, i.e., one color is removed from C; **(c)** $\bar{M}_{k-1,k-1}$ is colored with $k-2$ colors, i.e., two colors are removed from C.

*Study of case **(a)**: $\bar{M}_{k-1,k-1}$ remains colored with k colors*

By Item 3 of Lemma 5.5, v_k and u_k belong to two distinct horizontal colors, say S_1 ($\subset V$) and S_2 ($\subset U$), respectively. Assume now that $C = (S_1, \ldots, S_k)$ has m horizontal and $k-m$ vertical colors. Assume also that, up to a reordering of the colors, the m first ones are horizontal and the $k-m$ last ones are vertical; in other words, the vertical colors in C are $S_{i+1} = \{v_i, u_i\}$, $i = m, \ldots, k-1$ (both v_k and u_k belong to horizontal colors). Under this assumption, the graph induced by $S_1 \cup \ldots \cup S_m$, denoted by $\bar{M}'_1$, and the one induced by $S_{m+1} \cup \ldots \cup S_k$, denoted by $\bar{M}'_2$, are both bipartite complements of a perfect matching on vertex-sets $\{v_i, u_i : i = 1, \ldots, m-1, k\}$ and $\{v_i, u_i : i = m, \ldots, k-1\}$, respectively. Set $C'_1 = (S_1, \ldots, S_m)$ and $C'_2 = C \setminus C'_1 = (S_{m+1}, \ldots, S_k)$ and note that:

$$E\left(\bar{M}_{k,k}, C\right) = E\left(\bar{M}'_1, C'_1\right) + E\left(\bar{M}'_2, C'_2\right) \qquad [5.42]$$

3. As we have seen in section 5.4.3, such a graph is a bipartite graph $B(V, U, E)$ with $|V| = |U| = k$ and with $E = E(B_{k,k}) \setminus \{v_i u_i, v_i \in V, u_i \in U, i = 1, \ldots, k\}$, where by $B_{k,k}$ we denote the complete bipartite graph with $|V| = |U| = k$.

By Item 4 of Lemma 5.5, the first term of [5.42] is (strictly) greater than the value of the 2-coloring of M_1' that by the assumption of the proposition is (strictly) greater than $1-(1-p_k)^2+\sum_{i=1}^{m-1}(1-(1-p_i)^2)$. Consequently, the assumption of case **(a)** is in contradiction with the value of the coloring C assumed; in other words case **(a)** cannot occur under the hypothesis of the proposition. The proof of case **(a)** is now complete.

Study of case ***(b)****: $\bar{M}_{k-1,k-1}$ is colored with $k-1$ colors*

One color, say S_k, is removed from C with the removal of $\{v_k, u_k\}$. Denote by $C' = (S_1, \ldots, S_{k-1})$ the coloring so obtained. Two subcases may appear here: **(b.1)** both v_k and u_k belong to S_k and **(b.2)** v_k and u_k belong to two distinct colors.

Study of subcase ***(b.1)****: both v_k and u_k belong to S_k*

Since $\{v_k, u_k\}$ belong to the same color S_k, by Item 1 of Lemma 5.5 stated above, no other vertex can simultaneously belong to it. In this case, the value of C' is $\sum_{i=1}^{k-1}(1-(1-p_i)^2)$ and the proof of subcase **(b.1)** is complete.

Study of subcase ***(b.2)****: v_k and u_k belong to two distinct colors*

Assume that one among v_k, u_k, say v_k, belongs to a color in C', say S_1; then u_k is a color by itself, i.e., $S_k = \{u_k\}$. By Item 3 of Lemma 5.5, S_1 is horizontal. Assume as previously that coloring $C_1' = (S_1, \ldots, S_m)$ is horizontal (S_k is one among $S_1, \ldots, S_m$) and that $C_2' = (S_{m+1}, \ldots, S_{k-1})$ is vertical, denote by $\bar{M}_1'$ and $\bar{M}_2'$, the subgraphs of $\bar{M}_{k,k}$ induced by the vertex sets $S_1 \cup \ldots \cup S_m$ and $S_{m+1} \cup \ldots \cup S_{k-1}$, respectively, and note that both of them are complements of perfect bipartite matchings. Note also that [5.42] always holds. Then, exactly the same arguments as for case **(a)** conclude that C_1' is dominated by the 2-coloring of $\bar{M}_1'$ that is dominated by the (unique) vertical of this graph. Consequently, the assumption of subcase **(b.2)** is in contradiction with the value of C claimed in the proposition's statement; in other words, subcase **(b.2)** never occurs. The study of subcase **(b.2)** is now complete.

Study of case ***(c)****: $\bar{M}_{k-1,k-1}$ is colored with $k-2$ colors*

Here, v_k and u_k are two distinct colors by themselves, say S_{k-1} and S_k. Then, by application of Corollary 5.1, these colors can be mixed into one (vertical) color thus improving the functional which from $\sum_{i=1}^{k-2}(1-(1-p_i)^2)+2p_k$ now becomes (after the merging of the two colors) $\sum_{i=1}^{k}(1-(1-p_i)^2) < \sum_{i=1}^{k-2}(1-(1-p_i)^2)+2p_k$, contradicting so the assumption of the proposition on the value of coloring considered. Once more, case **(c)** cannot occur and its proof is complete.

In all we have proved that, under the assumptions made, only subcase **(b.1)** can hold. Moreover, from the proofs of the cases **(a)**, **(b)** and **(c)**, one can immediately deduce that the vertical coloring of $\bar{M}_{k,k}$ is the one with the smallest functional over any other feasible coloring of $\bar{M}_{k,k}$. An easy backwards induction finally shows that the claims of the proposition remain valid for any $k \geqslant 3$ and this completes its proof. ∎

A vertex-probability system satisfying Proposition 5.14 really exists. Indeed, the following lemma states that there exists a probability system for the vertices of a bipartite graph $\bar{M}_{k,k}$ such that, for any i, $3 \leqslant i < k$, the functional of a vertical coloring of any subgraph $\bar{M}_{i,i}$ of $\bar{M}_{k,k}$ is smaller than the functional of the 2-coloring of $\bar{M}_{i,i}$. The proof of this lemma can be found in section 5.9.5.

LEMMA 5.11.– *Consider a bipartite graph $\bar{M}_{n,n}$, set $V = \{v_1, \dots, v_n\}$ and $U = \{u_1, \dots, u_n\}$ both sets ranged in decreasing vertex-probability order. Set $p(v_i) = p(u_i) = \epsilon^i$, for $\epsilon \leqslant 1/3$. Then, this vertex-probability system verifies Proposition 5.14.*

Lemma 5.11 guarantees henceforth that Proposition 5.14 is feasible. We use it in order to face the main complexity result of this section, namely that PROBABILISTIC MIN k-COLORING(K) is **NP**-complete, for any $k > 3$.

THEOREM 5.9.– PROBABILISTIC MIN k-COLORING(K) *is **NP**-complete.*

Proof. Our problem is obviously in **NP**. On the other hand, in Proposition 5.13, we have proved that PROBABILISTIC MIN 3-COLORING(K) is **NP**-complete even for bipartite graphs, B' having the following three additional characteristics:

(a) only six vertices $v_i, u_i, i = 1, 2, 3$ of B' have non-zero probabilities;

(b) the subgraph of B' induced by these six vertices is a $\bar{M}_{3,3}$;

(c) $p_i = p(v_i) = p(u_i) = \epsilon^i$, $i = 1, 2, 3$, for a suitable ϵ, for example $\epsilon < 1/10$.

We reduce PROBABILISTIC MIN 3-COLORING(K') to PROBABILISTIC MIN k-COLORING(K). We consider an instance B' of the former, where B' fits characteristics (a) to (c) and $K' = \sum_{i=1}^{3}(1 - (1 - p_i)^2)$, and construct an instance B of the latter as follows: we put B' together with a $\bar{M}_{k-3,k-3}$; we link the vertices of $V(B')$ with the ones of $U(\bar{M}_{k-3,k-3})$ in such a way that the graph induced by $V(B') \cup U(\bar{M}_{k-3,k-3})$ is a complete bipartite graph; we do so with $U(B')$ and $V(\bar{M}_{k-3,k-3})$. We set $V(\bar{M}_{k-3,k-3}) = \{v_4, \dots v_k\}$, $U(\bar{M}_{k-3,k-3}) = \{u_4, \dots u_k\}$ and $p_i = p(v_i) = p(u_i) = \epsilon^i$, $i = 4, \dots, k$. Finally, we set $K = \sum_{i=1}^{k}(1 - (1 - p_i)^2)$.

By what has been discussed previously, around Proposition 5.14 and Item 1 of Lemma 5.5 (a vertical color on v_i and u_i cannot have vertices other than these two ones) and by the fact that the contribution of any other vertex of B' in the functional of any coloring is 0, one can immediately deduce that B has a k-coloring with functional at most $K = \sum_{i=1}^{k}(1 - (1 - p_i)^2)$, if and only if B' has a 3-coloring with functional at most $K' = \sum_{i=1}^{3}(1 - (1 - p_i)^2)$, QED. ∎

5.7.1.3. *Bipartite complements of bipartite matchings*

We now revisit bipartite complements of bipartite matchings, denoted by $\bar{M}_{n,n}$, already seen in section 5.4.3. We first make the following preliminary comment that indeed is a recall of Property 5.4.

REMARK 5.3.– An immediate consequence of Property 5.4 is that *the larger the large colors, the better the value of the coloring*. In other words, in the case of equal probabilities, good colorings are the ones where there exist some very small colors and some other very large ones. Let us call such colorings *unbalanced colorings* (in other words, an unbalanced coloring is a locally optimal coloring). If one could produce a k-coloring having, say, a very large color and $k-1$ singletons, this coloring would be the best among all the k-colorings of the graph, i.e., the best between the colorings of some size is the most unbalanced one. Conversely, the worst coloring is, in the case of identical probabilities, the most "balanced" one, i.e., the one where all the colors have the same size.

We are ready now to prove the following proposition.

PROPOSITION 5.15.– *Assuming identical vertex-probabilities,* PROBABILISTIC MIN k-COLORING *is polynomial in bipartite complements of bipartite matchings.*

Proof. Assume first that $k \geqslant n+1$ and set for facility $k = n + x$. Color $n - x + 1$ vertices, say, of U with the same color and then color the $n + x - 1$ still uncolored vertices using an unused color for any of them.

Assume now $k \leqslant n$. Color, say, U with one color. Color $n-(k-2)$ vertices of V with another color and then color the remaining $k-2$ vertices of V using an unused color for any of them. Following Remark 5.3 of section 5.7.3, in both of the cases discussed, the colorings produced are the most unbalanced ones, hence optimal. ∎

5.7.2. *The complements of bipartite graphs*

We now revisit the complements of bipartite graphs, seen in section 5.5. Note first that, assuming identical vertex probabilities, arguments very similar to those in section 5.5 show that in the complement $\bar{B}$ of any bipartite graph B of order n, $\chi(\bar{B}) = n - |M|$, where M is a maximum (cardinality) matching of B', where B' is as above. Dealing with complements of bipartite graphs, PROBABILISTIC MIN k-COLORING makes sense if $k > n - |M|$.

THEOREM 5.10.– *For any* $k > n - |M|$, PROBABILISTIC MIN k-COLORING *is polynomial in the complements of bipartite graphs, under any vertex-probability system.*

Proof. Note first that the discussion in section 5.5 exhibits the following facts dealing with any feasible k-coloring C of $\bar{B}$:

1) the 2-colors of $\bar{B}$ constitute a matching of B';

2) the number χ_2 of 2-colors of $\bar{B}$ verifies $\chi_2 = n - k$;

3) $E(\bar{B}, C) = \sum_{i=1}^{n} p_i - \sum_{\ell=1}^{\chi_2} p_{i_\ell} p_{j_\ell}$ (see also Lemma 5.9).

A combination of Facts 1 to 3 concludes that, for a given k, in order to determine the best k-coloring in $\bar{B}$, one has simply to compute the maximum-value matching (among those) of size $\chi_2 = n - k$ in B'.

Recall that computation of a maximum matching in a bipartite graph B' reduces to computation of a maximum integral flow in a transportation network N' (see Appendix A.2), derived from B' ([EVE 79]). Indeed, given $B'(U, V, E(B'))$, one adds two new vertices, a source s' and a sink t, one links s' to any vertex u of, say, U by an arc (s', u), and any vertex v in V to t by an arc (v, t) and transforms edges in $E(B')$ into arcs by orienting them from U to V. One sets lower bounds of all of the arcs to 0; capacities of arcs are set as follows: an arc of type (s', u) or (v, t) is assigned with capacity 1, while arcs derived from $E(B')$ are assigned with capacity ∞. Note also that, as one can see from the proof of [EVE 79], any feasible flow ϕ of the so constructed transportation network N' corresponds to a matching with size equal to the value $|\phi|$ of ϕ.

Now, in order to compute a matching of size, say m', one can modify N' by adding a new source s and by linking it to the former source s' by an arc (s, s') of capacity m' (and of lower bound 0). Let us denote by N the new network thus derived. Then, by arguments completely analogous to the ones in [EVE 79], the problem of computing a flow of maximum value m' in N amounts to computation of a maximum flow in N which, as observed above, allows computation of a maximum matching, i.e., of a matching of cardinality m' in the original bipartite graph B'.

However, our problem is to compute a maximum value matching among the ones of a fixed cardinality, say m' in an edge-valuated bipartite graph B'. Obviously, the construction of an edge-valued transportation network N can be performed as above by maintaining the same values on the arcs derived from $E(B')$ and by assigning value 0 on any other among the arcs added. Denoting by $\vec{w}$ the vector of the weights on the arcs of N, our problem then becomes:

$$\begin{cases} \max & \phi \cdot \vec{w} \\ \text{subject to} & \phi \text{ is a feasible flow of value } m' \text{ in } N \end{cases}$$

that, by very simple arguments of linear algebra, is equivalent to the following problem:

$$\begin{cases} \min & \phi \cdot (-\vec{w}) \\ \text{subject to} & \phi \text{ is a feasible flow of value } m' \text{ in } N \end{cases} \qquad [5.43]$$

Note now that Busacker's and Gowen's algorithm ([GON 85]), that works for weights of any sign, determines in polynomial time an optimal solution for the problem expressed by [5.43] (i.e., the one of determining a minimum-cost flow of a fixed value in a transportation network).

Consequently, given a bipartite graph B' (constructed from $\bar{B}$ as in the discussion before Theorem 5.4), one transforms it into a weighted one $(B', \vec{w})$ by weighting any edge (v_i, u_j) by $p_i p_j$. Then, one transforms $(B', \vec{w})$ into a weighted transportation network $(N, \vec{w}_N)$ as described above and then one applies the minimum-cost fixed value flow algorithm of Busacker and Gowen. By what has been discussed, this algorithm polynomially solves PROBABILISTIC MIN k-COLORING in complements of bipartite graphs, QED. ∎

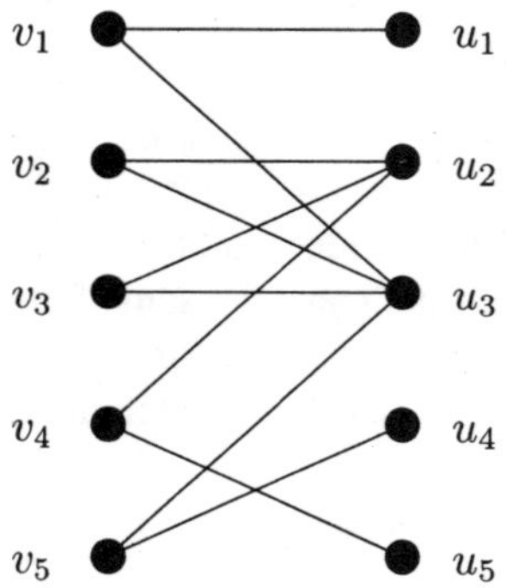

(a) A bipartite graph

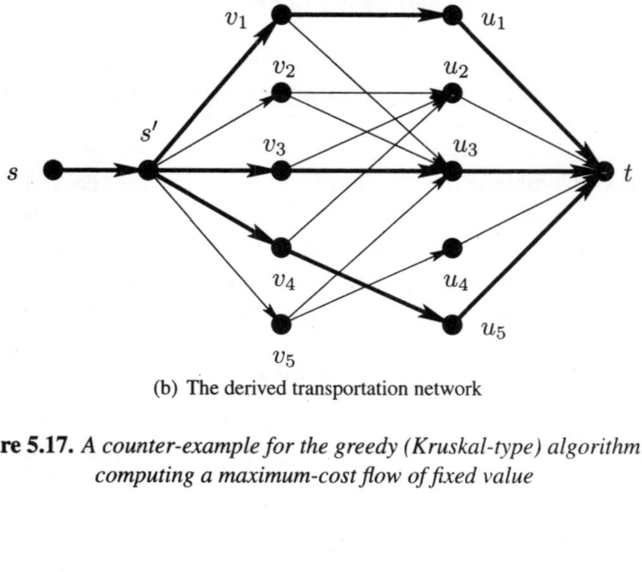

(b) The derived transportation network

Figure 5.17. *A counter-example for the greedy (Kruskal-type) algorithm for computing a maximum-cost flow of fixed value*

A question that should be addressed here is whether a simpler algorithm, for instance the greedy one, consisting of building a matching of size m' by using the m' heaviest edges that can form a matching (i.e., a strategy analogous to the one of Kruskal for minimum spanning tree ([KRU 56])), could be used to derive the result of Theorem 5.4. We show that this kind of procedure does not work for our problem. Consider the graph B' of Figure 5.17(a) with edge-weights: $w(v_1, u_1) = 100$, $w(v_1, u_3) = 1$, $w(v_2, u_2) = 2$, $w(v_2, u_3) = 9$, $w(v_3, u_2) = 9$, $w(v_3, u_3) = 10$, $w(v_4, u_2) = 1$, $w(v_4, u_5) = 100$, $w(v_5, u_3) = 2$, $w(v_5, u_4) = 1$, and the network N derived from it, shown in Figure 5.17(b) together with a flow of (maximum) value 210 and of size 3 (heavy-drawn arcs); recall that the capacity of the arc (s, s') is 3, for our case, and the weights on the additional arcs are all equal to 0. If one tries to construct a maximum-value flow of size 4, then the greedy algorithm would have used arcs (v_1, u_1), (v_4, u_5), (v_3, u_3) and (v_2, u_2) (resulting in a flow of size 4 and of value 212), while there exists a flow of size 4 with value 218. It is obtained by replacing arc (v_3, u_3) (of weight 10) in Figure 5.17(b) by the arcs (v_2, u_3) and (v_3, u_2) (both of weight 9).

On the contrary, such a simple algorithm is feasible when $\bar{B}$ has identical vertex-probabilities. In this case, all the colors of size 2 have the same value; the same holds, obviously, for the colors-singletons. Hence, all the colorings using χ_2 2-colors and χ_1 singleton-colors, i.e., $n - \chi_2$ colors, have the same functional's value. Suppose now that we ask for determining the best k-coloring of $\bar{B}$ for $k = n - |M| + x$ and for some $x \leqslant |M|$. Then, starting from a maximum matching M of B', one can split any x independent sets of size 2 into $2x$ singletons. This new configuration corresponds, with respect to $\bar{B}$, to a coloring of size $k = n - |M| + x$.

5.7.3. *Approximation in particular classes of graphs*

Let us restrict ourselves in graph-classes where the minimum coloring problem is polynomial and assume that the graphs considered have identical vertex probabilities, denoted by p.

Consider a k-colorable graph G, an optimal (deterministic) coloring $\hat{C}$ of G and denote by $\hat{k}$ its cardinality. Then, one can split some colors in order to produce a k-coloring. A possible way for doing this is, for example, to create $k - \hat{k}$ new singletons by "emptying" the smaller colors of $\hat{C}$ (note that, since we deal with identical probabilities, the smaller the color, the smaller its weight). Denote by C the so-obtained coloring and by C^* a minimum-functional one. Then, by Remark 5.3 (section 5.7.1.3), the following inequalities hold for $E(G, C)$ (in fact, for any k-coloring) and $E(G, C^*)$, respectively:

$$E(G, C) \leqslant \sum_{i=1}^{k} \left(1 - (1 - p)^{\frac{n}{k}}\right) \qquad [5.44]$$

$$E(G, C^*) \geqslant \left(1 - (1-p)^{n-k+1}\right) + (k-1)p \qquad [5.45]$$

Note that $n/k \leqslant n - k + 1$. Then, combination of [5.44] and [5.45] and some obvious algebra derives $E(G, C)/E(G, C^*) \leqslant 1/p$. So, the following corollary holds.

COROLLARY 5.10.– PROBABILISTIC MIN k-COLORING is approximable within approximation ratio $1/p$ in graphs with identical vertex-probabilities, where the minimum coloring problem is polynomial. In particular, if p is fixed, then PROBABILISTIC MIN k-COLORING belongs to **APX**.

Note that, using [5.4], an approximation ratio n/k is immediately derived for any k-colorable graph of order n with identical vertex-probabilities, where a k-coloring can be computed in polynomial time. Furthermore, the discussion just before [5.44] remains valid. So, the following corollary holds.

COROLLARY 5.11.– PROBABILISTIC MIN k-COLORING is approximable within approximation ratio n/k in graphs of order n with identical vertex-probabilities, where the minimum coloring problem is polynomial. In particular, if n/k is fixed, then PROBABILISTIC MIN k-COLORING is in **APX**.

Let us notice that the result of Corollary 5.11 identically holds for PROBABILISTIC MIN COLORING.

5.8. Comments and open problems

Table 5.1 summarizes the main results and open questions dealing with PROBABILISTIC MIN COLORING, arising from the chapter, while Table 5.2 does so for PROBABILISTIC MIN k-COLORING. Obviously, some of these results have several important corollaries (both tables are given at the end of the chapter). For instance, the fact that PROBABILISTIC MIN COLORING is polynomial in trees with bounded degrees and a fixed number of distinct probabilities has as a consequence that it is also polynomial in paths with a fixed number of distinct probabilities. Also, since PROBABILISTIC MIN COLORING is approximable within ratio 8/7 in general (i.e., under any system of vertex-probabilities) bipartite graphs, it is so in general trees, paths and even cycles.

What has been discussed in this chapter further confirms what we have claimed in Chapter 1 about the nature of probabilistic combinatorial optimisation problems. As we have quoted there, for these problems, even the simplest modification strategies may radically change their complexities with respect to their deterministic counterparts. PROBABILISTIC MIN COLORING is a very typical witness of this claim. In its deterministic version it is polynomial for both bipartite and chordal graphs, as well

Graph-classes	Complexity	Approximation ratio
General graphs	**NP**-hard	$O(n \log \log n / \log^{3/2} n)$
Bipartite graphs	?	8/7
Bipartite graphs, $p_i \geqslant 0.5$	Polynomial	
Trees	?	
Trees, bounded degree, k distinct probabilities	Polynomial	
Trees, all leaves exclusively at even or odd level, identical p_is	Polynomial	
Stars	Polynomial	
Paths	?	
Cycles	?	
Even cycles, odd cycles, paths, p_i identical	Polynomial	
Bipartite complements of bipartite matchings	Polynomial	
Complements of bipartite graphs	Polynomial	
Split graphs	**NP**-complete even for p_i identical	$1 + \epsilon$, for any $\epsilon > 0$

Table 5.1. *Summary of the main results of the chapter for* PROBABILISTIC MIN COLORING

Graph-classes	Complexity	Approximation ratio
Bipartite graphs	**NP**-hard, for any $k \geqslant 3$	?
Bipartite complements of bipartite matchings	?	?
Bipartite complements of bipartite matchings, p_i identical	Polynomial	
Complements of bipartite graphs	Polynomial	

Table 5.2. *Summary of the main results of the chapter for* PROBABILISTIC MIN k-COLORING

as for many other graph-classes (see [GAR 79] for more details). In its probabilistic version under the most simple, natural and intuitive modification strategy, the one considered in this chapter, it becomes hard for bipartite graphs. Furthermore, even if we restrict ourselves to the case of identical vertex-probabilities, no obvious technique seems to lead to polynomial results, even for general trees. On the other hand, if we assume distinct probabilities, the complexity of the problem remains unknown even in the simplest graph-structures such as chains or cycles.

This, to a large extent, is due to the functional associated with the strategy considered. While such a strategy applied to probabilistic independent set (Chapter 2), or to probabilistic vertex cover (Chapter 3), or even to probabilistic longest path (when measured in terms of vertices; Chapter 4) leads to weighted versions of the corresponding deterministic problems, this is not the case for PROBABILISTIC MIN COLORING. Here, as we have seen, the weight of an independent set is not an additive function and this ensures that MIN PROBABILISTIC COLORING becomes very particular and much harder than the probabilistic problems covered in the previous chapters. But such phenomena are precisely the beauty and the challenge of the probabilistic combinatorial optimization.

5.9. Proofs of the different results

5.9.1. *Proof of [5.5]*

For the left-hand side of [5.5], observe first that it is true for $\ell = 1$ and suppose it true for $\ell = \kappa$, i.e.:

$$\sum_{i=1}^{\kappa} p_i - \sum_{i=1}^{\kappa} \sum_{j=i+1}^{\kappa} p_i p_j \leqslant 1 - \prod_{i=1}^{\kappa} (1 - p_i) \qquad [5.46]$$

Suppose now that $\ell = \kappa + 1$. Expression [5.46] implies $\prod_{i=1}^{\kappa}(1 - p_i) \leqslant 1 - \sum_{i=1}^{\kappa} p_i + \sum_{i=1}^{\kappa} \sum_{j=i+1}^{\kappa} p_i p_j$. Multiply both terms of the last inequality by $(1 - p_{\kappa+1})$; then:

$$\begin{aligned}
\prod_{i=1}^{\kappa+1} (1 - p_i) &\leqslant \left(1 - \sum_{i=1}^{\kappa} p_i + \sum_{i=1}^{\kappa} \sum_{j=i+1}^{\kappa} p_i p_j\right)(1 - p_{\kappa+1}) \\
&= 1 - \sum_{i=1}^{\kappa} p_i + \sum_{i=1}^{\kappa} \sum_{j=i+1}^{\kappa} p_i p_j - p_{\kappa+1}
\end{aligned}$$

$$
\begin{aligned}
&+ p_{\kappa+1}\sum_{i=1}^{\kappa} p_i - p_{\kappa+1}\sum_{i=1}^{\kappa}\sum_{j=i+1}^{\kappa} p_i p_j \\
= &\ 1 - \sum_{i=1}^{\kappa+1} p_i + \sum_{i=1}^{\kappa+1}\sum_{j=i+1}^{\kappa+1} p_i p_j - p_{\kappa+1}\sum_{i=1}^{\kappa}\sum_{j=i+1}^{\kappa} p_i p_j \\
\leqslant &\ 1 - \sum_{i=1}^{\kappa+1} p_i + \sum_{i=1}^{\kappa+1}\sum_{j=i+1}^{\kappa+1} p_i p_j
\end{aligned}
$$

which proves the left-hand side inequality in [5.5]. For the right-hand side, we prove by induction on ℓ that $\prod_{i=1}^{\ell}(1 - p_i) \geqslant 1 - \sum_{i=1}^{\ell} p_i$. This is clearly true for $\ell = 1$; suppose it also true for $\ell = \kappa$, i.e., $\prod_{i=1}^{\kappa}(1-p_i) \geqslant 1-\sum_{i=1}^{\kappa} p_i$. Then, by multiplying both members of this inequality by $(1 - p_{\kappa+1})$, we find that the product obtained is equal to $1 - p_{\kappa+1} - \sum_{i=1}^{\kappa} p_i + p_{\kappa+1}\sum_{i=1}^{\kappa} p_i \geqslant 1 - \sum_{i=1}^{\kappa+1} p_i$, QED.

5.9.2. *Proof of [5.4]*

Consider again quantity $\prod_{v_i \in S_j}(1 - p_i)$ and note that simple mathematical arguments derive:

$$
\begin{aligned}
\prod_{v_i \in S_j}(1 - p_i) &= \exp\left\{\sum_{v_i \in S_j} \log(1 - p_i)\right\} \\
&= \exp\left\{-\sum_{v_i \in S_j} \log\left(\frac{1}{1 - p_i}\right)\right\}
\end{aligned}
\qquad [5.47]
$$

Consider function $f(x) = x - \log(1/(1 - x)) = x + \log(1 - x)$; it is decreasing with $x \in [0, 1]$ and, moreover, $f(0) = 0$. Therefore, $f(x) \leqslant 0$, in other words, $p_i \leqslant \log(1/(1 - p_i))$. Using this inequality together with [5.47], we get:

$$
\begin{aligned}
E(G, C) &= \sum_{j=1}^{k}\left(1 - \exp\left\{-\sum_{v_i \in S_j} \log\left(\frac{1}{1 - p_i}\right)\right\}\right) \\
&\geqslant \sum_{j=1}^{k}\left(1 - \exp\left\{-\sum_{v_i \in S_j} p_i\right\}\right)
\end{aligned}
\qquad [5.48]
$$

On the other hand, $\prod_{v_i \in S_j}(1-p_i) \leqslant (1-p_{\min})^{|S_j|} \leqslant 1-p_{\min}$. Summing for $j=1$ to k, we get:

$$E(G,C) = \sum_{j=1}^{k}\left(1 - \prod_{v_i \in S_j}(1-p_i)\right) \geqslant \sum_{j=1}^{k} p_{\min} = kp_{\min}$$

Note now that $E(G,C) = \sum_{j=1}^{k}(1-\prod_{v_i \in S_j}(1-p_i)) \leqslant k$. Furthermore, observe that $1-p_i \geqslant 1-p_{\max}$; hence, $\prod_{v_i \in S_j}(1-p_i) \geqslant (1-p_{\max})^{|S_j|}$. Let us prove that, for any $\ell > 0$, $(1-p_{\max})^{\ell} \geqslant 1-\ell p_{\max}$. This inequality is obviously true for $\ell = 1$. Suppose it is true for $\ell \leqslant \kappa$; in particular, for $\ell = \kappa$ we have: $(1-p_{\max})^{\kappa} \geqslant 1-\kappa p_{\max}$. At range $\kappa+1$ we have: $(1-p_{\max})^{\kappa+1} \geqslant (1-\kappa p_{\max})(1-p_{\max}) = 1-(\kappa+1)p_{\max} + \kappa p_{\max}^2 \geqslant 1-(\kappa+1)p_{\max}$ and the inequality claimed holds for any ℓ. Using it for $|S_j|$, $j = 1, \ldots, k$, we have $\prod_{v_i \in S_j}(1-p_i) \geqslant (1-p_{\max})^{|S_j|} \geqslant 1-|S_j|p_{\max}$, that implies $1-\prod_{v_i \in S_j}(1-p_i) \leqslant |S_j|p_{\max}$. Summing it for $j = 1, \ldots, k$, we get $E(G,C) = \sum_{j=1}^{k}(1-\prod_{v_i \in S_j}(1-p_i)) \leqslant \sum_{j=1}^{k}|S_j|p_{\max} = np_{\max}$.

5.9.3. *Proof of Property 5.1*

Between colorings C and C' the only colors changed are S_i and S_j. Then:

$$f(C') - f(C) = f(S_i') - f(S_i) + f(S_j') - f(S_j) \qquad [5.49]$$

Set now:

$$\begin{array}{rcccl} S_i' & = & (S_i \setminus \{x\}) \cup \{y\} & & \\ S_j' & = & (S_j \setminus \{y\}) \cup \{x\} & & \\ S_i'' & = & S_i \setminus \{x\} & = & S_i' \setminus \{y\} \\ S_j'' & = & S_j \setminus \{y\} & = & S_j' \setminus \{x\} \end{array} \qquad [5.50]$$

Then, using notations of [5.50], we get:

$$\begin{aligned} f(S_i') - f(S_i) = & \\ & 1-(1-p_y)\prod_{v_h \in S_i''}(1-p_h) - 1 + (1-p_x)\prod_{v_h \in S_i''}(1-p_h) \\ = & \ (p_y - p_x)\prod_{v_h \in S_i''}(1-p_h) \end{aligned} \qquad [5.51]$$

Similarly, we get:

$$f\left(S'_j\right) - f\left(S_j\right) = \left(p_x - p_y\right) \prod_{v_h \in S''_j} \left(1 - p_h\right) \qquad [5.52]$$

Using [5.51] and [5.52] in [5.49], we get:

$$f\left(C'\right) - f(C) = \left(p_y - p_x\right) \left(\prod_{v_h \in S''_i} \left(1 - p_h\right) - \prod_{v_h \in S''_j} \left(1 - p_h\right) \right) \qquad [5.53]$$

Recall that, by hypothesis, we have $f(S_i) \leqslant f(S_j)$ and $p_x \geqslant p_y$; consequently, by some easy algebra, we achieve $\prod_{v_h \in S''_i}(1 - p_h) - \prod_{v_h \in S''_j}(1 - p_h) \geqslant 0$ and, since $p_y - p_x \leqslant 0$, we conclude that the right-hand-side of [5.53] is negative, implying that coloring C' is better than C, QED.

5.9.4. *Proof of Proposition 5.2*

If $p_{\min} \geqslant 0.5$, then, for any color S_i of any coloring C' of B, $1 > f(S_i) \geqslant 0.5$. Hence, for any feasible coloring C' of B, $f(C') \geqslant 0.5|C'| > 0.5$. On the other hand, as $f(C) < 2$, the optimal coloring can never use more than 3 colors. So, in a first instance, an optimal coloring of B uses either 2 or 3 colors.

Consider any 3-coloring C' of B. Due to Properties 5.1 and 5.2, the best 3-coloring ever reachable (and possibly unfeasible) is coloring $C'' = (S''_1, S''_2, S''_3)$ assigning color S''_1 to a vertex of B with lowest probability (denote by v such a vertex), color S''_2 to a vertex with the second lowest probability (denote by $p'_{\min}$ this probability and by v' such a vertex) and color S''_3 to all the other vertices of B. It is easy to see that $f(S''_3) > f(S''_2) \geqslant f(S''_1)$. More precisely:

$$f\left(S''_1\right) = p_{\min} \qquad [5.54]$$

$$f\left(S''_2\right) = p'_{\min} \geqslant p_{\min} \qquad [5.55]$$

$$f\left(S''_3\right) \geqslant p'_{\min} \geqslant p_{\min}$$

Using [5.54] and [5.55] and the fact that $p_{\min} \geqslant 0.5$, we get:

$$f\left(S''_1\right) + f\left(S''_2\right) \geqslant 2p_{min} \geqslant 1 \qquad [5.56]$$

We will prove that $f(V) + f(U) \leqslant f(S_1'') + f(S_2'') + f(S_3'')$. For this, we distinguish the following four exhaustive cases, depending on the fact that v and v' belong to V, or to U:

1) $v \in V$ and $v' \in U$;

2) $v \in U$ and $v' \in V$;

3) $v, v' \in V$;

4) $v, v' \in U$.

We will examine Cases 1 and 3 as Case 2 is exactly specular to the former and Case 4 to the latter.

For Case 1, using [5.54], [5.55] and [5.56], one has to show that:

$$\begin{aligned} 1+1-\prod_{v_i \in (V \cup U)\setminus\{v,v'\}} (1-p_i) &= 2-\prod_{v_i \in (V \cup U)\setminus\{v,v'\}} (1-p_i) \\ &\geqslant 1-\prod_{v_i \in V}(1-p_i)+1-\prod_{v_i \in U}(1-p_i) \\ &= 2-\prod_{v_i \in V}(1-p_i)-\prod_{v_i \in U}(1-p_i) \end{aligned} \qquad [5.57]$$

or, equivalently:

$$\begin{aligned} &\prod_{v_i \in (V \cup U)\setminus\{v,v'\}} (1-p_i) - (1-p_{\min}) \prod_{v_i \in V\setminus\{v\}} (1-p_i) \\ &- (1-p'_{\min}) \prod_{v_i \in U\setminus\{v'\}} (1-p_i) \leqslant 0 \end{aligned} \qquad [5.58]$$

Set $\Gamma_1 = \prod_{v_i \in V\setminus\{v\}}(1-p_i)$ and $\Gamma_2 = \prod_{v_i \in U\setminus\{v'\}}(1-p_i)$. Then, [5.58] becomes:

$$\Gamma_1\Gamma_2 - (1-p_{\min})\Gamma_1 - (1-p'_{\min})\Gamma_2 \leqslant 0 \qquad [5.59]$$

Taking into account that $1 - p_{min} \geqslant \Gamma_1$ and $1 - p'_{min} \geqslant \Gamma_2$, [5.59] becomes $\Gamma_1^2 + \Gamma_2^2 - \Gamma_1\Gamma_2 = (\Gamma_1 - \Gamma_2)^2 + \Gamma_1\Gamma_2 \geqslant 0$, that is always true. The proof of Case 1 is complete.

We now analyze Case 3. By analogy with [5.58], we have to show that:

$$\prod_{v_i \in (V \cup U) \setminus \{v,v'\}} (1-p_i) - (1-p_{\min})(1-p'_{\min}) \prod_{v_i \in V \setminus \{v,v'\}} (1-p_i) - \prod_{v_i \in U} (1-p_i) \leqslant 0 \qquad [5.60]$$

Set this time $\Gamma_1 = \prod_{v_i \in V \setminus \{v,v'\}}(1-p_i)$ and $\Gamma_2 = \prod_{v_i \in U}(1-p_i)$. Then, [5.60] becomes:

$$\Gamma_1\Gamma_2 - (1-p_{\min})(1-p'_{\min})\Gamma_1 - \Gamma_2 \leqslant 0 \qquad [5.61]$$

or, equivalently $\Gamma_2(\Gamma_1 - 1) \leqslant (1-p_{\min})(1-p'_{\min})\Gamma_1$, which is always true since the left-hand quantity is negative and right-hand one is positive. This completes the proof of Case 3 and of the proposition.

5.9.5. *Proof of Lemma 5.11*

We fix a $k \leqslant n$ and we show that for any isomorphic $\bar{M}_{k,k}$ of $\bar{M}_{n,n}$, the functional of a vertical coloring of $\bar{M}_{k,k}$ is smaller than the functional of the 2-coloring $\bar{M}_{k,k}$. Order the vertices of $\bar{M}_{k,k}$ in decreasing order of probabilities and, without loss of generality, set $V_k = \{v_{k_1}, \ldots, v_{k_k}\}$ and $U_k = \{u_{k_1}, \ldots, u_{k_k}\}$. We will compute an ϵ such that, if we set $p_i = p(v_i) = p(u_i) = \epsilon^i$, $i = 1, \ldots n$, and if we assume a vertical coloring $C_k = \{\{v_{k_i}, u_{k_i}\} : i = 1, \ldots, k\}$ for $\bar{M}_{k,k}$, then:

$$\sum_{i=1}^{k} \left(1 - (1-p_{k_i})^2\right) \leqslant 2 - 2\prod_{i=1}^{k}(1-p_{k_i}) \qquad [5.62]$$

For the left-hand side of [5.62] we have: $\sum_{i=1}^{k}(1-(1-p_{k_i})^2) = 2\sum_{i=1}^{k} p_{k_i} - \sum_{i=1}^{k} p_{k_i}^2$. For the right-hand side of [5.62], we get, using the first inequality of [5.5]: $2 - 2\prod_{i=1}^{k}(1-p_{k_i}) \geqslant 2\sum_{i=1}^{k} p_{k_i} - 2\sum_{i=1}^{k}\sum_{j=i+1}^{k} p_{k_i}p_{k_j}$. Using this last expression, in order to prove [5.62], it suffices to compute p_{k_i} such that $2\sum_{i=1}^{k} p_{k_i} - \sum_{i=1}^{k} p_{k_i}^2 \leqslant 2\sum_{i=1}^{k} p_{k_i} - 2\sum_{i=1}^{k}\sum_{j=i+1}^{k} p_{k_i}p_{k_j}$, or:

$$\sum_{i=1}^{k} p_{k_i}^2 \geqslant 2\sum_{i=1}^{k} p_{k_i} \sum_{j=i+1}^{k} p_{k_j} \qquad [5.63]$$

Fix $k_i \in \{1, \ldots, k\}$; suppose that vertex $v_{k_i} \in V_k$ corresponds to vertex $v_\ell \in V$ ($\ell \geqslant k_i$) and recall that vertices are ranged in decreasing probability-order. We want to compute p_{k_i} (of the form ϵ^x for some integer $x \in 1, \ldots, n$), in such a way that [5.63] is satisfied, i.e., $p_{k_i} \geqslant 2\sum_{j=i+1}^{k} p_{k_j}$. Obviously, $2\sum_{j=i+1}^{k} p_{k_j} \leqslant 2\sum_{j=\ell+1}^{n} p_j$; therefore, we want to compute ϵ such that $p_{k_i} = \epsilon^\ell \geqslant 2\sum_{j=\ell+1}^{n} \epsilon^j$. The right-hand side of this inequality is a geometric series with ratio ϵ. Its value is $(\epsilon^{\ell+1} - \epsilon^{n+1})/(1-\epsilon)$; so, we compute ϵ such that $\epsilon^\ell \geqslant 2(\epsilon^{\ell+1} - \epsilon^{n+1})/(1-\epsilon)$, or $1 \geqslant 2(\epsilon - \epsilon^{n-\ell+1})/(1-\epsilon)$. Function $\ell \mapsto (\epsilon - \epsilon^{n-\ell+1})/(1-\epsilon)$ is decreasing with ℓ; so, $2(\epsilon - \epsilon^{n-\ell+1})/(1-\epsilon) \leqslant 2(\epsilon - \epsilon^n)/(1-\epsilon)$. So, we compute ϵ satisfying $1 \geqslant 2(\epsilon - \epsilon^n)/(1-\epsilon)$, i.e., $1 \geqslant 3\epsilon - 2\epsilon^n$ which is true for $\epsilon \leqslant 1/3$, QED.

SECOND PART

Structural Results

Chapter 6

Classification of Probabilistic Graph-problems

The goal of this chapter is to propose some structural results about probabilistic combinatorial optimization problems under strategies consisting either of moving absent elements out of the *a priori* solution (strategy MS, cf. Chapter 1), or, whenever this operation does not lead to feasible solutions, completing the result of the application of MS in order to render it feasible for the problems dealt.

Recall first that, when dealing with probabilistic combinatorial optimization, the main mathematical problems are the computational time of the functional's value, the characterization of the optimal *a priori* solution and the complexity of its computation. On the other hand, as we have seen in the previous chapters, when dealing with a probabilistic problem PΠ, considering MS leads to some particular weighted versions of the underlying deterministic optimization problem Π; in this sense, for such problems functional's computation is polynomial, and the optimal *a priori* solution and, consequently, the complexity of the probabilistic problem is well-characterized. Hence, it seems to us to be natural and interesting to give structural conditions characterizing these three main mathematical issues and to show that natural and well-known combinatorial problems fit these conditions.

6.1. When MS is feasible

In this section, we deal with graph-problems where a feasible solution is a subset of the vertex-set, or of the edge-set of the input-graph and where application of MS in the "present" (final) subgraph produces feasible solutions. Recall that, in the former case, MS consists, given an *a priori* solution S and a subgraph $G'(V', E') = G[V']$ of $G(V, E)$ induced by $V' \subseteq V$, of dropping absent vertices out of S, i.e., of taking $S' = S \cap V'$ as solution for G'. In the latter case, supposing that whenever a vertex

$v \in V$ is absent, so do all the edges incident to v. So, in this case, MS consists of taking $S' = S \cap E'$ as solution for G'.

6.1.1. *The a priori solution is a subset of the initial vertex-set*

The main result of this section is stated in Theorem 6.1. It gives sufficient conditions under which functionals are polynomially computed and *a priori* solutions are well-characterized.

THEOREM 6.1.– *Consider a probabilistic combinatorial optimization problem* PΠ *verifying the following assumptions:*

1) the instance of Π *is a vertex-weighted graph* $G(V, E, \vec{w})$ *(*$|V| = n$*); any vertex* $v_i \in V$ *has, with respect to* PΠ*, a presence-probability* p_i*;*

2) any feasible solution of Π *on any instance* G *is a subset of* V*;*

3) application of MS *in any of the* 2^n *subgraphs of* G *produces a feasible solution;*

4) the value of any solution $S \subseteq V$ *of* Π *is defined by:* $m(G, S) = w(S) = \sum_{v_i \in S} w_i$*, where* w_i *is the weight of* $v_i \in V$*.*

Then the functional of PΠ*, associated with* MS *is expressed as:*

$$E(G, S, \mathtt{MS}) = \sum_{v_i \in S} w_i p_i$$

and can be computed in polynomial time. Furthermore, the complexity of PΠ *is the same as the one of* Π*.*

Proof. Fix a subset $V' \subseteq V$ and an *a priori* solution S for PΠ on G. Denote by S' the solution for $G[V']$ computed by MS and set $G' = G[V']$. Following our hypotheses, S' is feasible for $G[V']$. Its value is given by:

$$m\left(G', S\left(G', \mathtt{MS}\right)\right) = \sum_{v_i \in S} w_i 1_{\{v_i \in V'\}} \qquad [6.1]$$

Using [6.1] in [1.3], we get for the functional:

$$\begin{aligned} E(G, S, \mathtt{MS}) &= \sum_{V' \subseteq V} m\left(G', S\left(G', \mathtt{MS}\right)\right) \Pr\left[G'\right] \\ &= \sum_{V' \subseteq V} \sum_{v_i \in S} w_i 1_{\{v_i \in V'\}} \Pr\left[V'\right] \\ &= \sum_{v_i \in S} w_i \sum_{V' \subseteq V} 1_{\{v_i \in V'\}} \Pr\left[V'\right] \qquad [6.2] \end{aligned}$$

For any vertex $v_i \in V$, set $\mathcal{V}_i' = \{V' \subseteq V : v_i \in V'\}$. Then, the last sum of [6.2] becomes:

$$\sum_{V' \subseteq V} 1_{\{v_i \in V'\}} \Pr[V'] = \sum_{V' \in \mathcal{V}_i'} \Pr[V'] \quad [6.3]$$

Set $V_i = V \setminus \{v_i\}$. Then, $\mathcal{V}_i' = \{V' \subseteq V : V' = \{v_i\} \cup V'', V'' \subseteq V_i\}$. Using also the fact that presence-probabilities of the vertices of V are independent (cf. Appendice A.3), we get from [6.3]:

$$\begin{aligned} \sum_{V' \subseteq V} 1_{\{v_i \in V'\}} \Pr[V'] &= \sum_{V'' \subseteq V_i} \Pr[\{v_i\} \cup V''] \\ &= \sum_{V'' \subseteq V_i} \Pr[v_i] \Pr[V''] \\ &= \Pr[v_i] \sum_{V'' \subseteq V_i} \Pr[V''] = p_i \end{aligned} \quad [6.4]$$

A combination of [6.2] and [6.4] immediately leads to the expression claimed for the functional.

It is easy to see that this functional can be computed in time linear in n. Furthermore, computation of the optimal *a priori* solution for PΠ in G obviously amounts to the computation of the optimal weighted solution for Π in $G(V, E, \vec{w}')$, where, for any $v_i \in V$, $w_i' = w_i p_i$. Consequently, by this observation and by assumption 4 in the statement of the theorem, Π and PΠ have the same complexity. ∎

"Why, although computation of the functional is *a priori* exponential (since it carries over 2^n subgraphs of G), assumptions of Theorem 6.1 allow polynomial computation of its value?" Because, under these assumptions, given a subgraph G' induced by a subset $V' \subseteq V$, the value of the solution for G' is the sum of the weights of the vertices in $S \cap V'$. Furthermore, a vertex not in S will never make part of any solution in any subgraph of G. Consequently, computation of the functional amounts to determining, for any G', which vertices make part of $S \cap V'$. This is equivalent to the specification, for any $v_i \in S$, of all the subgraphs to which v_i belongs and to a summation of the presence-probabilities of these subgraphs. This sum is, as we have seen in [6.4], equal to p_i, i.e., the probability of v_i. This simplification is the reason that renders functional's computation polynomial.

Theorem 6.1 has the following immediate corollary dealing with the case of probabilistic versions of unweighted combinatorial optimization problems.

COROLLARY 6.1.– Consider a probabilistic combinatorial optimization problem PΠ verifying the following assumptions:

1) the instance of Π is a graph $G(V, E)$ ($|V| = n$); any vertex $v_i \in V$ has, with respect to PΠ, a presence-probability p_i;

2) any feasible solution of Π on any instance G is a subset of V;

3) application of MS in any of the 2^n subgraphs of G produces a feasible solution;

4) the value of any solution $S \subseteq V$ of Π is defined by: $m(G, S) = |S|$.

Then, the functional of PΠ, associated with MS, is expressed as:

$$E(G, S, \texttt{MS}) = \sum_{v_i \in S} p_i$$

and can be computed in polynomial time. Furthermore, PΠ is equivalent to a weighted version of Π where vertex-weights are the vertex-probabilities.

Notice that Corollary 6.1 is somewhat weaker than Theorem 6.1 since it does not establish the equivalence between Π and PΠ. Indeed, this result can be seen as a kind of reduction from Π to PΠ stating that the latter is *a priori* harder than the former one.

It is easy to see that PROBABILISTIC MAX INDEPENDENT SET and PROBABILISTIC MIN VERTEX COVER under modification strategy MS (cf., also, Chapters 2 and 3, respectively) fit conditions of Theorem 6.1 and Corollary 6.1.

We now focus ourselves in the case of VERTEX LONGEST PATH and VERTEX WEIGHTED LONGEST PATH. Recall first that the strategy MS for these problems is exactly the strategy called MV in Chapter 4. Both problems clearly fit assumptions 1, 2, 4 and 1, 2, 4 of Theorem 6.1 and Corollary 6.1, respectively. Dealing with assumptions 3, things are somewhat more complicated. If no further hypothesis on the form of the input-graph is made, then these two assumptions are not always satisfied. Indeed, denote by L_V the list of vertices in an *a priori* solution S and suppose that v_i, v_j and v_k are three consecutive vertices in L_V. Suppose furthermore that v_j is absent from G'. Then, following strategy MS, arcs (v_i, v_j) and (v_j, v_k) (being absent themselves) should be replaced by (v_i, v_k). If this arc does not exist in G, then application of MS is not feasible. On the other hand, if one makes the additional assumption that G is transitive, then arc (v_i, v_k) exists and, consequently, MS becomes feasible and, in this case, both Theorem 6.1 and Corollary 6.1 apply. In other words, PROBABILISTIC VERTEX WEIGHTED LONGEST PATH and PROBABILISTIC VERTEX LONGEST PATH, restricted in transitive connected acyclic directed graphs, fit all the assumptions of Theorem 6.1 and Corollary 6.1, respectively.

PROPOSITION 6.1.– *Consider modification strategy MS.*

1) Given an instance $G(V, E, \vec{w})$ and an independent set S for MAX WEIGHTED INDEPENDENT SET *(resp.,* MAX INDEPENDENT SET*), the functional associated with S and* MS *is $E(G, S, \mathtt{MS}) = \sum_{v_i \in S} w_i p_i$ (resp., $E(G, S, \mathtt{MS}) = \sum_{v_i \in S} p_i$). The optimal a priori solution S^* is then a maximum-weight independent set in the graph $G(V, E, \vec{w}')$, where $\vec{w}'$ is such that $w_i' = w_i p_i$ (resp., p_i), $i = 1, \ldots, n$.*

2) Given an instance $G(V, E, \vec{w})$ and a vertex cover C for MIN WEIGHTED VERTEX COVER *(resp.,* MIN VERTEX COVER*), the functional associated with C and* MS *is $E(G, C, \mathtt{MS}) = \sum_{v_i \in C} w_i p_i$ (resp., $E(G, C, \mathtt{MS}) = \sum_{v_i \in C} p_i$). The optimal a priori solution C^* is then a minimum-weight vertex cover in the graph $G(V, E, \vec{w}')$, where $\vec{w}'$ is such that $w_i' = w_i p_i$ (resp., p_i), $i = 1, \ldots, n$.*

3) Given an instance $G(V \cup \{s, t\}, A, \vec{w})$, representing a vertex-weighted transitive connected acyclic oriented graph and a path L_V (as a list of consecutive vertices) for VERTEX WEIGHTED LONGEST PATH *(resp.,* VERTEX LONGEST PATH*), the functional associated with L_V and* MS *is $E(G, L_V, \mathtt{MS}) = \sum_{v_i \in L_V} w_i p_i$ (resp., $E(G, L_V, \mathtt{MS}) = \sum_{v_i \in L_V} p_i$). The optimal a priori solution L_V^* is then a vertex-weight longest path (i.e., a path maximizing the sum of the weights of its vertices) in the graph $G(V \cup \{s, t\}, A, \vec{w}')$, where $\vec{w}'$ is such that $w_i' = w_i p_i$ (resp., p_i), $i = 1, \ldots, n$.*

6.1.2. *The a priori solution is a collection of subsets of the initial vertex-set*

We now deal with problems the feasible solutions of which are collections of subsets of the initial vertex-set. Consider a graph $G(V, E)$ and a (deterministic) combinatorial optimization graph-problem Π whose set of feasible solutions are collections of subsets of V verifying some specified non-trivial hereditary property[1] (e.g., independent set, clique, etc.). The following theorem characterizes functionals and optimal *a priori* solutions for such problems.

THEOREM 6.2.– *Consider a probabilistic combinatorial optimization problem* PΠ *verifying the following assumptions:*

1) the instance of Π is a graph $G(V, E)$ (with $|V| = n$); any vertex $v_i \in V$ has, with respect to PΠ*, a presence-probability p_i;*

2) any feasible solution of Π on any instance G is a collection $C = (V_1, \ldots, V_k)$ of subsets of V any of them satisfying some specified non-trivial hereditary property;

3) application of MS *in any of the 2^n subgraphs of G produces a feasible solution;*

4) the value of any solution S of Π is defined by: $m(G, S) = k$ (the size of the collection C).

1. A property π is *hereditary* if, whenever is satisfied by a graph G', it is satisfied by any subgraph of G'; a hereditary property π is non-trivial if it is true (satisfied) for infinitely many graphs and false for infinitely many graphs.

Then, the functional of PΠ, *associated with* MS *is expressed as:*

$$E(G, S, \texttt{MS}) = \sum_{j=1}^{k} \left(1 - \prod_{v_i \in V_j} (1 - p_i)\right)$$

and can be computed in polynomial time. The probabilistic counterpart PΠ *of* Π *under* MS*, amounts at a particular weighted version of* Π*, where the weight of any vertex* $v_i \in V$ *is* $1 - p_i$*, the weight* $w(V_j)$ *of a subset* $V_j \subseteq V$ *is defined by* $w(V_j) = \prod_{v_i \in V_j}(1 - p_i)$ *and the objective function to be optimized is equal to* $\sum_{V_j \in C}(1 - w(V_j))$.

Proof. Consider an *a priori* solution $C = (V_1, V_2, \ldots, V_k)$ of cardinality k and a subgraph $G' = G[V']$ of G. Denote by $k' = m(G', C(G', \texttt{MS}))$ the value of the solution returned by MS on G'. As previously, denote by $\mathbf{1}_{\mathsf{F}}$ the indicator function of a fact F. Recall that:

$$E(G, C, \texttt{MS}) = \sum_{V' \subseteq V} \Pr[V']\, m(G', C(G', \texttt{MS})) = \sum_{V' \subseteq V} \Pr[V']\, k' \qquad [6.5]$$

Consider the facts:

– $\mathsf{F_j}$: subset $\mathsf{V_j}$ has at least one vertex;

– $\bar{\mathsf{F}}_\mathsf{j}$: application of MS leaves subset $\mathsf{V_j}$ empty;

then k' can be written as $k' = \sum_{j=1}^{k} \mathbf{1}_{\mathsf{F_j}} = \sum_{j=1}^{k}(1 - \mathbf{1}_{\bar{\mathsf{F}}_\mathsf{j}})$ and [6.5] becomes:

$$\begin{aligned}
E(G, C, \texttt{MS}) &= \sum_{V' \subseteq V} \Pr[V'] \left(\sum_{j=1}^{k} \left(1 - \mathbf{1}_{\bar{\mathsf{F}}_\mathsf{j}}\right)\right) \\
&= \sum_{V' \subseteq V} \Pr[V'] \sum_{j=1}^{k} 1 - \sum_{V' \subseteq V} \Pr[V'] \sum_{j=1}^{k} \mathbf{1}_{\mathsf{V_j \cap V' = \emptyset}} \\
&= \sum_{j=1}^{k} \sum_{V' \subseteq V} \Pr[V'] - \sum_{j=1}^{k} \sum_{V' \subseteq V} \Pr[V']\, \mathbf{1}_{\mathsf{V_j \cap V' = \emptyset}} \\
&= k - \sum_{j=1}^{k} \prod_{v_i \in V_j} (1 - p_i) \\
&= \sum_{j=1}^{k} \left(1 - \prod_{v_i \in V_j} (1 - p_i)\right) \qquad [6.6]
\end{aligned}$$

It is easy to see that computation of $E(G, C, \texttt{MS})$ can be performed in at most $O(n)$ steps; consequently, PΠ is in **NPO**. Furthermore, by [6.6], the characterization of the feasible solutions for PΠ claimed in the statement of the theorem is immediate. ∎

What does play a central role for yielding result of Theorem 6.2 is the fact that the the sets of the collection C satisfy some mathematical property which is hereditary. This makes it possible to preserve sets $V_1, \ldots, V_k$ in the solution returned by MS unless they are empty and, consequently, to express $E(G, C, \texttt{MS})$ as in [6.6], in terms of facts $\mathsf{F_j}$ and $\bar{\mathsf{F}}_\mathsf{j}$.

Assume that $p_i = 1$, for any $v_i \in V$. Then, by [6.6], $E(G, C, \texttt{MS}) = k$, i.e., PΠ coincides with Π and Theorem 6.2 has the following immediate corollary.

COROLLARY 6.2.– If Π is **NP**-hard, then PΠ is so.

It is easy to see that MIN COLORING under modification strategy MS (cf., also, Chapter 5) fits conditions of Theorem 6.2 and Corollary 6.2, respectively. In Chapter 7, we will see some other problems also fitting the same conditions.

6.1.3. *The a priori solution is a subset of the initial edge-set*

We deal in this section with problems for which feasible solutions are sets of edges and where MS, consisting of moving absent edges[2] out from an *a priori* solution S, is feasible. In other words, the basic hypothesis in this section is that given an instance $G(V, E)$ of a problem Π having as feasible solutions subsets of E, an *a priori* solution $S \subseteq E$ and a subgraph $G'(V', E') = G[V']$, $V' \subseteq V$, the set $S' = S \cap E'$ is a feasible solution for Π when dealt in G'. The main result for this case is the following theorem.

THEOREM 6.3.– *Consider a probabilistic combinatorial optimization problem* PΠ *verifying the following assumptions:*

1) the instance of Π *is an edge- (or arc-) valued graph* $G(V, E, \vec{\ell})$ ($|V| = n$)*; any vertex* $v_i \in V$ *has, with respect to* PΠ*, a presence-probability* p_i*;*

2) any feasible solution of Π *on any instance* G *is a subset of* E*;*

3) application of MS *in any of the* 2^n *subgraphs of* G *produces a feasible solution;*

4) the value of any solution $S \subseteq E$ *of* Π *is defined by:*

$$m(G, S) = w(S) = \sum_{(v_i, v_j) \in S} \ell\left(v_i, v_j\right)$$

where $\ell(v_i, v_j)$ *is the valuation of the edge (or arc) of* (v_i, v_j) *of* G*.*

2. Recall that lack of a vertex entails lack of any edge incident to it.

Then, the functional of PΠ, *associated with* MS *is expressed as:*

$$E(G, S, \mathtt{MS}) = \sum_{(v_i,v_j)\in S} \ell(v_i, v_j)\, p_i p_j$$

and can be computed in polynomial time. Furthermore, dealing with their respective computational complexities, PΠ *and* Π *are equivalent.*

Proof. Set $S' = S \cap E'$; by the assumptions of the theorem, S' is feasible for Π when dealing with G'. Furthermore:

$$m(G', S(G', \mathtt{MS})) = \sum_{(v_i,v_j)\in S} \ell(v_i, v_j)\, 1_{\{(v_i,v_j)\in E'\}} \qquad [6.7]$$

Using [6.7] in [1.3], we get for the functional:

$$\begin{aligned} E(G, S, \mathtt{MS}) &= \sum_{V'\subseteq V} m(G', S(G', \mathtt{MS})) \Pr[V'] \\ &= \sum_{V'\subseteq V} \sum_{(v_i,v_j)\in S} \ell(v_i, v_j)\, 1_{\{(v_i,v_j)\in E'\}} \Pr[V'] \\ &= \sum_{(v_i,v_j)\in S} \ell(v_i, v_j) \sum_{V'\subseteq V} 1_{\{(v_i,v_j)\in E'\}} \Pr[V'] \end{aligned} \qquad [6.8]$$

Any edge (or arc) $(v_i, v_j) \in E$ belongs to $G' = G[V']$, if and only if both of its endpoints belong to V'. We denote by $\mathcal{V}'_{ij} = \{V' \subseteq V : v_i \in V', v_j \in V'\}$ the set of all the subsets of V containing both v_i and v_j. It holds that:

$$\sum_{V'\subseteq V} 1_{\{(v_i,v_j)\in E'\}} \Pr[V'] = \sum_{V'\in\mathcal{V}'_{ij}} \Pr[V'] \qquad [6.9]$$

Set $V_{ij} = V \setminus \{v_i, v_j\}$. Then, $\mathcal{V}'_{ij} = \{V' \subseteq V : V' = \{v_i\} \cup \{v_j\} \cup V'', V'' \subseteq V_{ij}\}$. Using also the fact that presence-probabilities of the vertices of V are independent (cf. Appendice A.3), we get from [6.9]:

$$\sum_{V'\subseteq V} 1_{\{(v_i,v_j)\in E'\}} \Pr[V'] = \sum_{V''\subseteq V_{ij}} \Pr[\{v_i\} \cup \{v_j\} \cup V'']$$

$$= \sum_{V'' \subseteq V_{ij}} \Pr[v_i] \Pr[v_j] \Pr[V'']$$

$$= \Pr[v_i] \Pr[v_j] \sum_{V'' \subseteq V_{ij}} \Pr[V'']$$

$$= p_i p_j \qquad [6.10]$$

A combination of [6.8] and [6.10] immediately leads to the expression claimed for the functional.

It is easy to see that this functional can be computed in time quadratic in n. Furthermore, computation of the optimal *a priori* solution for PΠ in G obviously amounts to the computation of the optimal solution for Π in an edge- (or arc-) valued graph $G(V, E, \vec{\ell'})$ where, for any $(v_i, v_j) \in E$, $\ell'(v_i, v_j) = \ell(v_i, v_j) p_i p_j$. Consequently, by this observation and by assumption 4 in the statement of the theorem, Π and PΠ have the same complexity. ∎

The reasons for which the functional deduced in Theorem 6.3 are quite analogous to the ones in Theorem 6.1. Indeed, since an edge that does not belong to the *a priori* solution S will never be part of any solution in any subgraph $G'(V', E')$ of G, the computation of the functional amounts to the specification, for any G', of the set $S \cap E'$. This is equivalent to first determining, for any edge $e \in S$, all the subgraphs containing e and next to a summation of the probabilities of these subgraphs.This sum equals to the product of the probabilities of the endpoints of e.

As in section 6.1.1 (for Theorem 6.1), the following immediate corollary dealing with the case of probabilistic versions of unitary edge- (or arc-) valued combinatorial optimization problems can be deduced from Theorem 6.3.

COROLLARY 6.3.– Consider a probabilistic combinatorial optimization problem PΠ verifying the following assumptions:

1) the instance of Π is a graph $G(V, E)$ ($|V| = n$); any vertex $v_i \in V$ has, with respect to PΠ, a presence-probability p_i;

2) any feasible solution of Π on any instance G is a subset of E;

3) application of MS in any of the 2^n subgraphs of G produces a feasible solution;

4) the value of any solution $S \subseteq E$ of Π is defined by: $m(G, S) = |S|$.

Then, the functional of PΠ, associated with MS is expressed as:

$$E(G, S, \texttt{MS}) = \sum_{(v_i, v_j) \in S} p_i p_j$$

and can be computed in polynomial time. Furthermore, PΠ is equivalent to an edge- (or arc-) valued version of Π where the values of an edge is the product of the probabilities of its endpoints.

Let us note that, for instance, PROBABILISTIC MAX MATCHING in both edge-valued and non-valued graphs fits conditions of Theorem 6.3 and Corollary 6.3. Moreover, since MAX WEIGHTED MATCHING is polynomial, both PROBABILISTIC MAX WEIGHTED MATCHING and PROBABILISTIC MAX MATCHING are polynomial also.

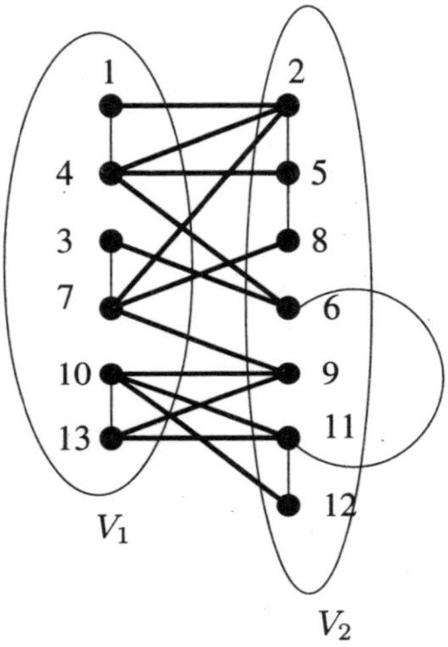

Figure 6.1. *A graph G with a cut S (thick edges)*

We now deal with MAX CUT in both edge-valued and unitary edge-valued graphs. Lets us first note that we can represent an *a priori* cut S as a set of edges in such a way that whenever $(v_i, v_j) \in S$, $v_i \in V_1$ and $v_j \in V_2$.

For example, in Figure 6.1, where for simplicity values of edges are not mentioned, the cut partitions V in subsets $V_1 = \{1, 3, 4, 7, 10, 13\}$ and $V_2 = \{2, 5, 6, 8, 9, 11, 12\}$ and the *a priori* cut S (thick edges) can then be written as

$$S = \{(1,2), (3,6), (4,2), (4,5), (4,6), \ldots, (13,11)\}$$

(where edges are ranged in lexicographic order). In Figure 6.2, we present graph's and cut's states assuming that vertices 4, 6 and 11 are absent. The result of strategy MS is a solution S' missing in all edges of S having at least one endpoint among $\{4, 6, 11\}$. In this figure also, one can see that, after application of MS, some vertices may exist, such as vertex 3 in Figure 6.2. that have no incident cut-edges.

Hence, both weighted and cardinality PROBABILISTIC MAX CUT meet the conditions of Theorem 6.3 and Corollary 6.3, respectively. Consequently, MAX CUT being **NP**-hard, so are PROBABILISTIC MAX WEIGHTED CUT and PROBABILISTIC MAX CUT.

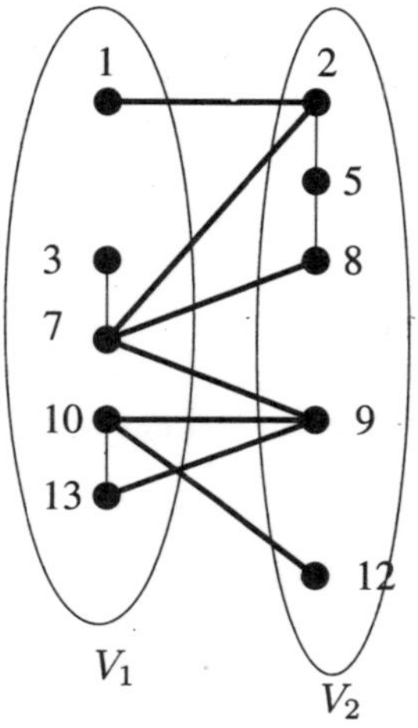

Figure 6.2. *Application of* MS *on the surviving subgraph and the a priori solution of Figure 6.1*

PROPOSITION 6.2.– *Consider modification strategy* MS.

1) Given an edge-valued instance $G(V, E, \vec{\ell})$ *and a matching* M *for* MAX WEIGHTED MATCHING *(resp.,* MAX MATCHING*), the functional associated with* M *and* MS *is:*

$$E(G, M, \mathtt{MS}) = \sum_{(v_i, v_j) \in M} \ell(v_i, v_j)\, p_i p_j$$

(resp., $E(G, M, \mathtt{MS}) = \sum_{(v_i,v_j)\in M} p_i p_j$*). The optimal a priori solution* M^* *is then a maximum-weight matching in the graph* $G(V, E, \vec{\ell'})$*, where* $\vec{\ell'}$ *is such that* $\ell'(v_i, v_j) = \ell(v_i, v_j) p_i p_j$ *(resp.,* $p_i p_j$*),* $i, j = 1, \ldots, |E|$*. Both* PROBABILISTIC MAX MATCHING *and* PROBABILISTIC MAX WEIGHTED MATCHING *belong to* ***P****.*

2) Given an instance $G(V, E, \vec{\ell})$ *and a cut* C *for* MAX WEIGHTED CUT *(resp.,* MAX CUT*), the functional associated with* C *and* MS *is:*

$$E(G, C, \mathtt{MS}) = \sum_{(v_i, v_j) \in C} \ell(v_i, v_j)\, p_i p_j$$

(resp., $E(G, C, \mathtt{MS}) = \sum_{(v_i,v_j)\in C} p_i p_j$*). The optimal a priori solution* C^* *is then a maximum-weight cut in the graph* $G(V, E, \vec{\ell'})$*, where* $\vec{\ell'}$ *is such that* $\ell'(v_i, v_j) =$

$\ell(v_i, v_j)p_ip_j$ *(resp.,* p_ip_j*),* $i, j = 1, \ldots, |E|$. MAX CUT *being* ***NP****-hard, so are* PROBABILISTIC MAX WEIGHTED CUT *and* PROBABILISTIC MAX CUT.

6.2. When application of MS itself does not lead to feasible solutions

We now deal with problems where application of MS does not immediately provide feasible solutions for any subgraph of the input graph. We focus ourselves on graph-problems where feasible solutions are connected sets of edges (for example, paths, trees, cycles, etc.) and where one can render the result of MS feasible by adding some new edges (present in the subinstance at hand).

Consider a problem Π where a feasible solution is a connected set S of edges. Consider also that vertices in S are ordered in some appropriate order. Assume that application of strategy MS results in a set of connected subsets $C_1, C_2, \ldots, C_k$ of S but that $S'' = \cup_{i=1}^{k} C_i$ is not connected (i.e., S'' does not constitute a feasible solution for Π). We consider a kind of "completion" of MS by additional edges linking, for $i = 1, \ldots, k-1$, the last vertex (in the ordering considered for C) of C_i with the first vertex of C_{i+1}. In other words, the new strategy denoted MSC considered in this section is the following (assuming an *a priori* solution S representing a connected set of edges):

1) arrange the vertices of S following some appropriate order;

2) apply strategy MS; let $C_1, C_2, \ldots, C_k$ be the connected components of S resulting from the application of MS;

3) for $i = 1, \ldots, k-1$, use an edge to link the last vertex of C_i with the first vertex of C_{i+1};

4) output S' the solution so computed.

In what follows, we denote by $V[S']$ the set of vertices in S' and by $G''(V[S'], E'')$ the graph $G[V[S']]$.

6.2.1. *The functional associated with* MSC

We show in this section that there exist a class of probabilistic combinatorial optimization problems for which the functional associated with MSC can be computed in polynomial time.

THEOREM 6.4.– *Consider a probabilistic combinatorial optimization problem* PΠ *verifying the following assumptions:*

1) instances of Π *are edge-valued complete graphs* $(K_n, \vec{\ell}) = G(V, E, \vec{\ell})$*; furthermore, in the probabilistic version of* Π *any vertex* $v_i \in V$ *has a presence-probability* p_i*;*

2) a feasible solution of Π *is a subset* S *of* E *verifying some connectivity property;*

3) given an a priori solution S *(the vertices of which are arranged in some appropriate order), strategy* `MSC` *computes a feasible solution* S'*, for any subgraph* $G'(V', E', \vec{\ell}) = G[V']$ *of* G *(obviously,* G' *is complete);*

4) $m(G, S) = \sum_{(v_i,v_j)\in S} \ell(v_i, v_j)$.

Denote by $[v_i, v_j]$ *the set* $\{v_{i+1}, v_{i+2}, \ldots, v_{j-1}\}$ *(*$i < j$ *in the ordering assumed for* S*) such that, for any* $k = i, i+1, \ldots, j-1$*,* $(v_k, v_{k+1}) \in S$ *(i.e.,* $[v_i, v_j]$ *is the set of vertices in the path linking* v_i *to* v_j *in* S*, where* v_i *and* v_j *themselves are not encountered). Then, the functional* $E(G, S, \texttt{MSC})$ *for* $\text{P}\Pi$ *is computable in polynomial time and is expressed by:*

$$E(G,S,\texttt{MSC}) = \sum_{(v_i,v_j)\in S} \ell(v_i,v_j)\, p_i p_j + \sum_{(v_i,v_j)\in E''\setminus S} \ell(v_i,v_j)\, p_i p_j \prod_{v_l\in[v_i,v_j]} (1-p_l)$$

where E'' *and* S *are as defined previously. Also, it is assumed that if* $[v_i, v_j] = \emptyset$*, then* $\prod_{v_l\in[v_i,v_j]}(1-p_l) = 0$. The proof of Theorem 6.4 is given in section 6.4.

The fact that $E(G, S, \texttt{MSC})$ is polynomial is partly due to the same reasons as in Theorems 6.1 and 6.3. Furthermore, the order of the additional edges-choices in Step 3 of `MSC` are also crucial for this efficiency. Indeed, this order is such that one can say *a priori* under which conditions an edge (or arc) (v_i, v_j) will be added in S'. These conditions carry over the presence or the absence of the edges initially lying before (v_i, v_j) in S.

Unfortunately, in the opposite of Theorems 6.1 and 6.3, Theorem 6.4 does not derive a "good" characterization of the optimal *a priori* solution of the problems meeting the assumptions stated. In particular, the form of the functional does not imply solution of some well-defined weighted version of Π (the deterministic support of $\text{P}\Pi$). This is due to the fact that in the expression of Theorem 6.4 for the functional, the "costs" assigned to the edges depend on the structure of the *a priori* solution chosen.

6.2.2. *Applications*

Let us first recall that section 4.2 of Chapter 4 deals with the hypotheses of Theorem 6.4. In what follows, we outline some other cases of problems where strategy `MS` does not lead to feasible solutions. In particular, we deal with problems where feasible solutions are either cycles or trees.

6.2.2.1. *The a priori solution is a cycle*

We deal in this section with one of the most paradigmatic combinatorial optimization problems, the MIN TSP and its probabilistic version, called PROBABILISTIC MIN TSP. We shall see any Hamiltonian cycle T (i.e., a feasible solution for MIN TSP, also called a tour in what follows) as a set of edges; its value is $m(K_n, T) = \sum_{e \in T} \ell(v_i, v_j)$. Moreover, we arbitrarily number the vertices of K_n in the order that they are visited in T; so, we can set

$$T = \{(v_1, v_2), \ldots, (v_i, v_{i+1}), \ldots, (v_{n-1}, v_n), (v_n, v_1)\}$$

If we consider an *a priori* tour T and a set of absent vertices, then application of MS may result in a set $\{P_1, P_2, \ldots, P_k\}$ of paths[3], ranged in the order that vertices have been visited in T, that is not feasible for MIN TSP in the surviving graph. Then, in order to render this set feasible, one can link (modulo k) the last vertex of the path P_i to the first vertex of P_{i+1}; this is always possible since the initial graph is complete. The tour thus obtained is denoted by T'.

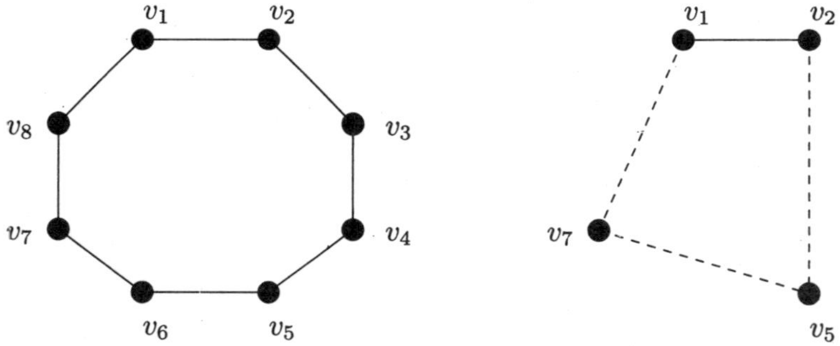

(a) An *a priori* tour (b) The tour T'

Figure 6.3. *An example of application of strategy MSC for* PROBABILISTIC MIN TSP

For example, in Figure 6.3(a), an *a priori* cycle T derived from a (symmetric) K_8 is shown. In Figure 6.3(b), we consider that vertices v_3, v_4, v_6 and v_8 are absent. On a first occasion, application of MS results in a path-set $\{\{(v_1, v_2)\}, \{v_5\}, \{v_7\}\}$. On a second occasion, we will link vertex v_2 to v_5 (using dotted edge (v_2, v_5)) and vertex v_5 to v_7 (by dotted edge (v_5, v_7)). This creates a Hamiltonian path linking all the surviving vertices of the initial K_8. Finally, we link vertex v_7 to v_1 (by dotted

3. These paths may be sets of edges, or simple edges, or even isolated vertices, any such vertex considered as a path.

edge (v_7, v_1). We thus build a new tour feasibly visiting all the present vertices of the remaining graph.

It is easy to see that for the cases we deal with, all conditions of Theorem 6.4 are verified. Consequently, its application for the case of PROBABILISTIC MIN TSP gives:

$$E(K_n, T, \texttt{MSC}) = \sum_{(v_i,v_j)\in T} \ell(v_i, v_j)\, p_i p_j + \sum_{(v_i,v_j)\in E(K'_n)\setminus T} \ell(v_i, v_j)\, p_i p_j \prod_{v_l \in [v_i,v_j]} (1 - p_l) \qquad [6.11]$$

where K'_n denotes the surviving (complete) subgraph of K_n after removal of the absent vertices, and $E(K'_n)$ its edge-set. We recover here the result of [JAI 85] about PROBABILISTIC MIN TSP. We see from [6.11] that the *a priori* solution minimizing the functional cannot be characterized tightly, since the expression for $E(K_n, T, \texttt{MSC})$ depends on the particular *a priori* tour T considered and by the way in which this particular tour will be completed in the surviving instance. The same expressions as in [6.11] for the functionals and the same corollaries about the optimal *a priori* solutions minimizing them also hold for the two most notorious restrictive cases of MIN TSP, the METRIC MIN TSP and the GEOMETRIC MIN TSP (called also MIN EUCLIDEAN TSP). For the former version, edge distances are considered to verify the triangle inequalities[4], while for the latter version, the points are considered in the plane and, for any two points $c_i = (x_i, y_i)$ and $c_j = (x_j, y_j)$, the distance between (x_i, y_i) and (x_j, y_j) is the discretized Euclidean length $\sqrt{(x_i - x_j)^2 + (y_i - y_j)^2}$.

6.2.2.2. *The a priori solution is a tree*

Let us consider here another very well-known combinatorial optimization problem, the MIN SPANNING TREE. For reasons that will be understood later, we restrict ourselves to complete graphs (see also [BER 90a]). As mentioned previously in section 6.2.2.1, we consider a tree by the set of its edges. For any tree T its value is $m(K_n, T) = \sum_{e\in T} \ell(v_i, v_j)$.

Let us note that either for MIN TSP dealt in section 6.2.2.1, or for the longest path dealt in section 4.2 of Chapter 4, their solutions induced an implicit and natural ordering of the edges of their solutions. This is not the case here since various orderings can be considered. We consider the one given by a kind of inverse depth-first-search ([AHO 75, BAA 78]) and obtained as follows:

4. Informally, for any pair of points, it is shorter to go from one to the other one directly than to pass from a third point, that is, for any triple (c_i, c_j, c_k), $d(c, c') \leqslant d(c, c'') + d(c'', c')$, where c, c' and c'', stand for any of c_i, c_j and c_k.

– starting from leaf numbered 1, we first number its incident edge, then one of the edges intersecting it and so on, until we arrive at a second leaf;

– we then return back to the last vertex of the path incident to an edge not numbered yet and we continue the numbering from this edge until a third leaf is encountered;

– we continue in this way until all the edges of T are visited.

This ordering is performed in $O(n)$ for a tree of order n (recall that such a tree has $n-1$ edges).

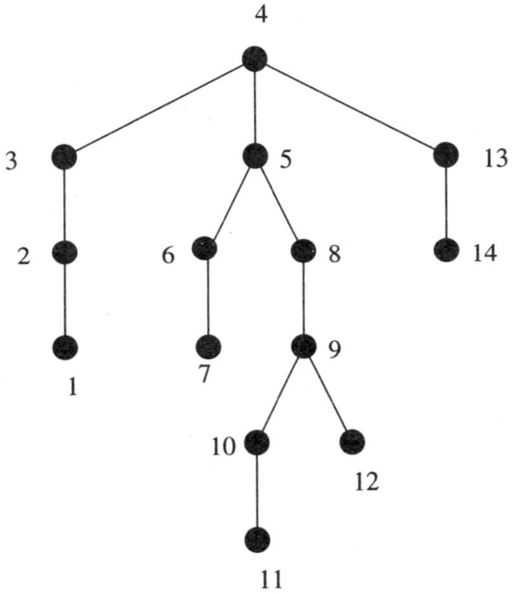

Figure 6.4. *The ordering of the edges of a tree together with the ordering of its vertices*

For example, consider the tree (of order 14) of Figure 6.4 and assume that it is a minimum spanning tree of sum graph of order 14. Starting from the leftmost leaf (numbered 1), we visit edges $(1,2), (2,3), \ldots, (6,7)$ in this order; node 7 is another leaf. We then return to the node 5 that is the last vertex incident to an edge not yet visited, and we visit edges $(5,8), (8,9)(9,10), (10,11)$. We then return to the node numbered 9 and visit edge $(9,12)$. We return back to the node numbered 4 and visit edges $(4,13), (13,14)$ in this order.

Let us note that the ordering exhibited above partitions the edges of the tree into a number of edge-disjoint paths $P_1, P_2, \ldots$ For instance, dealing with Figure 6.4, the depicted tree T is partitioned into 4 paths: $P_1 = \{(1,2),(2,3),\ldots,(6,7)\}$, $P_2 = \{(5,8),(8,9)(9,10),(10,11)\}$, $P_3 = \{(9,12)\}$ and $P_4 = \{(4,13),(13,14)\}$.

Suppose now that some vertices are absent from the initial graph G. Also, let us refer to the nodes by their numbering in the ordering just described. Then, strategy MS will produce a non-connected set of edges (forming paths, any of them being a subset of some P_i; denote by $\{P'_1, P'_2, \ldots, P'_k\}$ the set of paths obtained after application of MS). Order these paths following the order their edges appear in T. As done previously, in section 6.2.2.1, for any $l = 1, \ldots, k$, we link the last vertex, say i of path P'_l, to the first vertex, say j, of the path P'_{l+1}. Since the initial graph is assumed to be complete, such an edge always exists.

We have now to specify the path $[i,j]$ associated with the edge (i,j) connecting P'_l and P'_{l+1}. Starting from T, the edges of which are ordered as described, one can, without loss of generality, renumber the vertices of the graph in such a way that T can be rewritten as $T = \{(1,2),(2,3),\ldots,(n-k,n)\}$. Then, T can be seen as a sequence of vertices, some of them appearing more than once, i.e., $T = (1_1, 2_1, 3_1, \ldots, j_1, i_2, (j+1)_1, \ldots, (n-k)_q, n_1)$, where $i < j$ and i_c represents the c-th time where vertex i appears in T. Based upon this representation, one can reconstruct T in the following way: for any pair $(i_c, j_{c'})$ of consecutive vertices, edge (i,j) belongs to T if and only if $i < j$. Note that a leaf appears only once in the list and that the number of the vertex succeeding it in the list is smaller than the one of the leaf.

Application of strategy MS implies that, whenever an absent vertex v is not a leaf, its removal disconnects T. In order to render it connected, we have to link the vertex preceding it in the sequence to the one succeeding it. By the particular form of the sequence considered, any edge (i,j) that strategy MSC is likely to add in order to reconnect T is such that $i < j$; the corresponding path $[i,j]$ (i.e., the list of vertices that have to be absent in order that (i,j) is added) is the portion of the list between i_l and j_1, where i_l is the last occurrence of i before the first occurrence j_1 of vertex j.

Let us revisit the example of Figure 6.4. The sequence associated with the tree is $T = (1_1, \ldots, 7_1, 5_2, 8_1, \ldots, 11_1, 9_2, 12_1, 4_2, 13_1, 14_1)$. Assume now that vertices 2, 5, 11 and 13 disappear from the initial graph. Application of MS results in the subsequence $T' = (1_1, 3_1, 4_1, 6_1, 7_1, 8_1, 9_1, 10_1, 9_2, 12_1, 4_2, 14_1)$. The first connected component (path) in this sequence is vertex 1 itself, and the second one is vertices 3 and 4; we thus add edge $(1,3)$ in order to connect these two components. The third component is induced by vertices 6 and 7; edge $(4,6)$ is consequently added. The fourth connected component is induced by vertices 8, 9, 10 and 12; hence, edge $(7,8)$ is added. The fifth connected component in the sequence is vertex 4 itself but $(12,4)$ is not added since $12 > 4$. Finally, the sixth connected component, vertex 14, entails

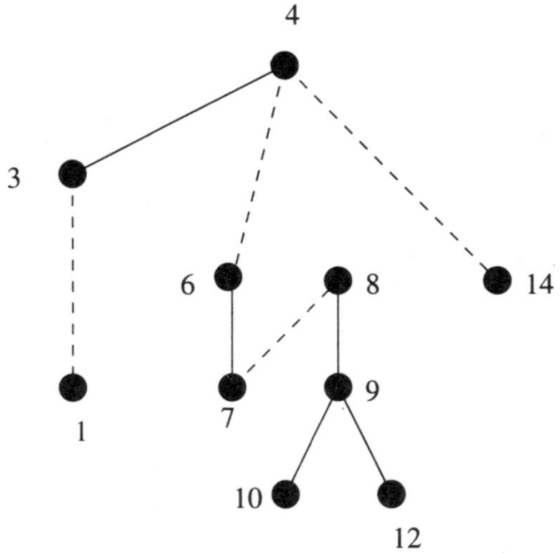

Figure 6.5. *The solution T' derived from application of strategy* MSC *in the a priori tree T of Figure 6.4*

introduction of edge $(4, 14)$ that completes the modification of T by strategy MSC. Figure 6.5, where slotted edges represent edges added during execution of Step 3 of MSC, illustrates what has been just discussed.

Denoting also by T the sequence of vertices representing the *a priori* spanning tree T, denoting any vertex of the initial graph by its numbering in the inverse dfs, explained earlier, and by T' the tree resulting from the application of MSC on T, and setting $[i, j] = \{q \in T : q \text{ a vertex between } i_l \text{ and } j_1 \text{ in } T\}$, the application of Theorem 6.4 leads to the following functional for PROBABILISTIC MIN SPANNING TREE under strategy MSC:

$$E\left(K_n, T, \texttt{MSC}\right) = \sum_{(i,j) \in T} \ell\left(i, j\right) p_i p_j + \sum_{(i,j) \in E(K'_n) \setminus T} \ell\left(i, j\right) p_i p_j \prod_{l \in [i,j]} \left(1 - p_l\right)$$

Another interesting version of MIN TSP is the so called MIN METRIC BOTTLENECK WANDERING SALESPERSON PROBLEM. In this problem, we are given a set C of n points $c_1, \ldots, c_n$ one among them considered as initial point, denoted by s, and another one considered as final point, denoted by f, together with distances $d(c_i, c_j)$ between any pair (c_i, c_j) of points. The distances between any pair of points satisfy triangle inequalities. The objective is to determine simple path from the initial point s

to the final point f passing through all the other points in C, only once[5], i.e., a permutation $\sigma : \{1, \dots, n\} \to \{1, \dots, n\}$, with $c_{\sigma(1)} = s$ and $c_{\sigma(n)} = f$, minimizing length of the largest distance in the path, i.e., the quantity $\max_{i=1,\dots,n-1}\{d(c_{\sigma(i)}, c_{\sigma(i+1)})\}$.

Obviously, the set of the feasible solutions for this problem is the same as for MIN TSP. Supposing that s and f are always present in the final subgraph, i.e., that their presence probabilities are equal to 1, it is easy to deduce that strategy MSC is feasible for MIN METRIC BOTTLENECK WANDERING SALESPERSON. Denoting by P an *a priori* Hamiltonian path, we get:

$$\begin{aligned} E\left(K_n, P, \texttt{MSC}\right) &= \sum_{V' \subseteq V} \Pr\left[V'\right] m\left(K_n\left[V'\right], P\left(V', \texttt{MSC}\right)\right) \\ &= \sum_{V' \subseteq V} \Pr\left[V'\right] \max_{i=1,\dots,n-1} \left\{d\left(c_{\sigma(i)}, c_{\sigma(i+1)}\right)\right\} \\ &\quad \times 1_{\left\{c_{\sigma(i)} \text{ present and } c_{\sigma(i+1)} \text{ present}\right\}} \end{aligned} \qquad [6.12]$$

From [6.12], computation of $E(K_n, P, \texttt{MSC})$ does not seem polynomial. Furthermore, an *a priori* solution minimizing it is not tightly characterizable.

6.3. Some comments

The major conclusion that seems to be brought out from what has been studied in this chapter deals with the complexity of the probabilistic combinatorial optimization problems. Indeed, it seems that when direct application of MS leads to feasible solutions, the complexity of determining the optimal *a priori* solution for the problems covered amounts to the complexity of solving some weighted version of the deterministic problem, where the weights depend on the vertex-probabilities. Moreover, these weights seem not to depend on particular characteristics of the *a priori* solution considered. On the contrary, when more-than-one-stage strategies are necessary for building solutions of the probabilistic counterpart of the problem dealt, then the observation above is no longer valid. Indeed, it seems that one also recovers some weighted version of the original deterministic problem, but the weights on the data cannot be assigned independently of the structure of a particular *a priori* solution.

Another comment deals with the framework adopted in this chapter. We have tried to reduce studies about both the complexity of the probabilistic problems and

5. In other words, we are asked to determine a Hamiltonian path with specified extremities s and t.

of the computation of their functionals to verifications of hypotheses about combinatorial characteristics of their deterministic counterparts. For instance, for some of the problems studied, the complexity of functional's computation simply amounts to establishing a certain order on the elements of the *a priori* solution.

In Table 6.1, a summary of the results about the problems considered in this chapter is given. In the first column we briefly recall the conditions (characteristics) of the solutions of their deterministic counterparts and the type of the modification strategy used. In the second column, the complexity of the functional is given and, when this is possible, the complexity of the probabilistic problem itself. Finally, in the third column, problems dealing with the conditions in the first column are cited.

6.4. Proof of Theorem 6.4

Denote by $C[E']$ the set of edges added to S'' during the execution of Step 3 of MSC. Obviously, $S' = S'' \cup C[E']$; also, if an edge belongs to $C[E']$, then it necessarily belongs to $E[S]$, the set of edges in E induced by the endpoints of the edges in S. By Assumptions 1 to 3, S' is a feasible set of edges. Furthermore:

$$\begin{aligned} m\left(G', S\left(G', \mathtt{MSC}\right)\right) &= \sum_{(v_i,v_j)\in E} \ell\left(v_i, v_j\right) 1_{\{(v_i,v_j)\in S'\}} \\ &= \sum_{(v_i,v_j)\in E} \ell\left(v_i, v_j\right) 1_{\{(v_i,v_j)\in S''\cup C[E']\}} \end{aligned} \quad [6.13]$$

By construction, any element of $C[E']$ is an edge (or arc), the initial endpoint of which corresponds to the terminal endpoint of a connected subset C_i of S and the terminal endpoint of which corresponds to the initial endpoint of the "next" connected C_{i+1} of S. Then, for any subgraph G' of G, the following two assertions hold:

– $S' \subseteq E''$;

– any edge that does not belong to E'', will never be part of any feasible solution; indeed, for such an edge, at least one of its endpoints does not belong to $V[S']$; so, $C[E'] \subseteq E''$.

We so have from [6.13], setting $S' = S(G', \mathtt{MSC})$:

$$m\left(G', S'\right) = \sum_{(v_i,v_j)\in E} \ell\left(v_i, v_j\right) 1_{\{(v_i,v_j)\in S''\cup C[E']\}}$$

Hypotheses	General results	Particular problems
Solution: some vertex-subset Solution's value: total weight MS feasible (Theorem 6.1)	Functional in $O(n)$ Identical complexities of Π and Probabilistic Π	PROBABILISTIC MAX WEIGHTED INDEPENDENT SET PROBABILISTIC MIN WEIGHTED VERTEX COVER PROBABILISTIC VERTEX WEIGHTED LONGEST PATH
Solution: some edge-subset Solution's value: total weight MS feasible (Theorem 6.3)	Functional in $O(n^2)$ Identical complexities of Π and Probabilistic Π	MAX WEIGHTED MATCHING MAX WEIGHTED CUT
Solution: some edge-subset Solution's value: total weight MSC feasible (but not MS) (Theorem 6.4)	Functional in $O(n^3)$	PROBABILISTIC ARC WEIGHTED LONGEST PATH PROBABILISTIC MIN TSP PROBABILISTIC MIN SPANNING TREE

Table 6.1. *The main results of the chapter.*

$$
\begin{aligned}
&= \sum_{(v_i,v_j)\in E''} \ell(v_i,v_j)\, 1_{\{(v_i,v_j)\in S''\cup C[E']\}} \\
&= \sum_{(v_i,v_j)\in E''} \ell(v_i,v_j)\, 1_{\{(v_i,v_j)\in S''\}} \\
&\quad + \sum_{(v_i,v_j)\in E''} \ell(v_i,v_j)\, 1_{\{(v_i,v_j)\in C[E']\}} \\
&= \sum_{(v_i,v_j)\in S} \ell(v_i,v_j)\, 1_{\{(v_i,v_j)\in E'\}} \\
&\quad + \sum_{(v_i,v_j)\in E''\setminus S} \ell(v_i,v_j)\, 1_{\{(v_i,v_j)\in C[E']\}} \qquad [6.14]
\end{aligned}
$$

A combination of [1.3] and [6.14] gives:

$$
\begin{aligned}
E(G,S,\mathtt{MSC}) &= \sum_{V'\subseteq V} \left(\sum_{(v_i,v_j)\in S} \ell(v_i,v_j)\, 1_{\{(v_i,v_j)\in E'\}} \right. \\
&\quad \left. + \sum_{(v_i,v_j)\in E''\setminus S} \ell(v_i,v_j)\, 1_{\{(v_i,v_j)\in C[E']\}} \right) \Pr[V'] \\
&= \sum_{V'\subseteq V} \sum_{(v_i,v_j)\in S} \ell(v_i,v_j)\, 1_{\{(v_i,v_j)\in E'\}} \Pr[V'] \\
&\quad + \sum_{V'\subseteq V} \sum_{(v_i,v_j)\in E''\setminus S} \ell(v_i,v_j)\, 1_{\{(v_i,v_j)\in C[E']\}} \Pr[V'] \\
&= \sum_{(v_i,v_j)\in S} \ell(v_i,v_j) \sum_{V'\subseteq V} 1_{\{(v_i,v_j)\in E'\}} \Pr[V'] \\
&\quad + \left(\sum_{(v_i,v_j)\in E''\setminus S} \ell(v_i,v_j) \right. \\
&\quad \left. \times \sum_{V'\subseteq V} 1_{\{(v_i,v_j)\in C[E']\}} \Pr[V'] \right) \qquad [6.15]
\end{aligned}
$$

As in the proof of Theorem 6.3, the first term of [6.15] can be simplified as follows:

$$\sum_{(v_i,v_j)\in S} \ell\left(v_i, v_j\right) \sum_{V'\subseteq V} 1_{\{(v_i,v_j)\in E'\}} \Pr\left[V'\right] = \sum_{(v_i,v_j)\in S} \ell\left(v_i, v_j\right) p_i p_j \qquad [6.16]$$

Using [6.16] in [6.15], we get:

$$E\left(G, S, \texttt{MSC}\right) = \sum_{(v_i,v_j)\in S} \ell\left(v_i, v_j\right) p_i p_j + \sum_{(v_i,v_j)\in E''\setminus S} \ell\left(v_i, v_j\right) \sum_{V'\subseteq V} 1_{\{(v_i,v_j)\in C[E']\}} \Pr\left[V'\right] \qquad [6.17]$$

We now deal with the second term of [6.17] that, in this form, seems to be exponential. Consider some edge (v_i, v_j) added during Step 3 in order to "patch", say, connected components C_l and C_{l+1} of the *a priori* solution S. Since $(v_i, v_j) \notin S$, there exists in S a sequence $\mu = [v_i, v_j]$ of consecutive edges (or arcs) linking v_i to v_j. Suppose that this sequence is listed by its vertices and that neither v_i nor v_j belong to μ. Edge $(v_i, v_j) \in E'' \setminus S'$ is added to S' just because all the vertices in μ are absent. In other words, inclusion $(v_i, v_j) \in C[E']$ holds for any subgraph $G'(V', E')$, where $V' \in \mathcal{U}'_{ij}$ with:

$$\mathcal{U}'_{ij} = \{V' \subseteq V : v_i \in V', v_j \in V', \text{ any vertex of } \mu \text{ is absent}\}$$

Consequently, the second sum in the second term of [6.17] can be written as:

$$\begin{aligned}\sum_{V'\subseteq V} 1_{\{(v_i,v_j)\in C[E']\}} \Pr\left[V'\right] &= \sum_{V'\in\mathcal{U}'_{ij}} \Pr\left[V'\right] \\ &= p_i p_j \prod_{v_l\in[v_i,v_j]} \left(1 - p_l\right) \end{aligned} \qquad [6.18]$$

A combination of [6.15], [6.17] and [6.18] derives the expression claimed for the functional in the statement of the theorem. It is easy to see that computation of a single term in the second sum of the functional requires $O(n)$ computations (at most $n+1$ multiplications). Since we may do this for at most $O(n^2)$ times (the edges in E), it follows that the whole complexity of functional's computation is of $O(n^2)$, which concludes the proof of the theorem.

Chapter 7

A Compendium of Probabilistic **NPO** Problems on Graphs

As has been understood in the previous chapters, the basic problematic of probabilistic combinatorial optimization can be briefly described as follows. A generic instance I of a combinatorial optimization problem Π is given. One assumes that Π is not to be necessarily solved on the whole I, but rather on a (unknown *a priori*) subinstance $I' \subset I$. Suppose that any datum d_i in the data-set describing I has a probability p_i, indicating how d_i is likely to be present in the final subinstance I'. One has to compute an *a priori* solution S for Π in the entire instance I and once I' becomes known, to move elements of S that do not belong to I' out from S (providing that this deletion results in a feasible solution for I') returning so a solution S' fitting I'. The objective is to determine an initial solution S for I such that, for any subinstance $I' \subseteq I$ presented for optimization, the solution S' obtained as described above respects some pre-defined quality criterion (for example, optimal for I', or achieving, say, constant approximation ratio, etc.). The modification strategy just outlined is strategy denoted by MS in the previous chapters. Note also that any of the particular graph-problems considered in Chapters 2 to 5 are first studied under this strategy. Finally, observe that as has been seen in Chapter 6, such a strategy is not always feasible, in the sense that it does not always lead to feasible solutions for I'.

In this chapter, we present a list of the most known and well-studied graph-problems and we investigate if strategy MS is feasible for any of them. For the problems for which it is feasible, we express the functional associated with it, and, when possible, we characterize the optimal *a priori* solution and the complexity of its computation. Whenever no comment about the computation of functional's value is made, this computation is polynomial, i.e., the corresponding probabilistic problem is in **NPO**. All the problems studied here (except MAX MATCHING) are classified either in [GAR 79], or in the Compendium of [AUS 99]. We have mainly been inspired by the classification of [AUS 99]. There, the most well-known and natural **NPO** graph-problems are classified into two sections including ten main categories, namely:

1) Graph theory
   a) covering and partitioning,
   b) subgraphs and supergraphs,
   c) vertex ordering,
   d) iso- and other morphisms,
   e) miscellaneous;
2) Network design
   a) spanning trees,
   b) cuts and connectivity,
   c) routing problems,
   d) flow problems,
   e) miscellaneous.

In this chapter, we follow the categorization of [AUS 99] by mixing sections "Graph theory" and "Network design" and by omitting categories where for any of the problems included strategy MS is not feasible. For instance:

– for Category 1c), vertex ordering, MS is unfeasible for any of the problems included; this is due to the fact that feasible solutions for these problems are one-to-one functions f (see section A.1 in Appendix A) from V (the vertex-set of the input-graph) to the set $\{1, 2, \ldots, |V|\}$; the absence of some vertices will produce a surviving vertex-set V' and then function $f : V' \rightarrow \{1, 2, \ldots, |V|\}$ is not one-to-one anymore;

– also, in Category 1e), the feasibility conditions of the problems studied deal with the existence of structures, such as paths or trees, that verify connectivity properties that make strategy MS unfeasible;

– the same observation also holds for the problems of Category 2a);

– in Category 2d), the connectivity properties (induced by the structures of the paths through which pass the feasible flows) that have to be satisfied by the feasible solutions make MS unfeasible unless particular properties (such as transitivity, etc.) are admitted for the input graphs;

– finally, in Category 2e), since solutions verify specific cardinality-conditions, the absence of some vertices might violate them; so, MS is no more feasible for problems in this category.

Note also that, for economy, we do not deal in the sequel with routing problems. Indeed, from the list of routing problems in [AUS 99], MS (completed into MSC) is feasible only for MIN TSP, MIN METRIC TSP, MIN GEOMETRIC TSP and MIN METRIC BOTTLENECK WANDERING SALESPERSON (for this last problem we have, in addition, to assume that s and f have presence-probabilities equal to 1). For the other ones, as they are defined in general graphs and particular connectivity conditions are required to be verified by a feasible solution, neither strategy MS nor its completion into MSC is feasible, unless particular forms (complete, or transitive, etc.) are assumed for the input-graphs.

In all, problems of the following categories are studied:

– covering and partitioning,

– subgraphs and supergraphs,

– iso- and other morphisms,

– cuts and connectivity.

For simplification, since throughout this chapter we only deal with a single strategy MS, we use notation $E(G, S)$, instead of $E(G, S, \mathtt{MS})$, for the functional associated with an input-graph G, an *a priori* solution S and MS. Also, in order to avoid redundancies, the following conventions are adopted in what follows:

– When mentioned that a problem[1] fits conditions of Theorem 6.1 in Chapter 6, then:

- its functional, associated with an *a priori* solution V', is expressed as:

$$E(G, V') = \sum_{v_i \in V'} p_i$$

- its probabilistic version becomes a weighted version where a weight equal to its presence-probability is associated with any vertex of V and where the objective becomes to optimize the total weight of a solution (i.e., the sum of the weights of the vertices in V');

- since setting $p_i = 1$, for any $v_i \in V$, the deterministic problem and its probabilistic counterpart coincide, the complexity of the latter is the same as the former.

– When mentioned that a problem[2] fits conditions of Theorem 6.2 in Chapter 6, then:

- its functional associated with an *a priori* solution C is expressed as:

$$E(G, C) = \sum_{i=1}^{k}\Bigl(1 - \prod_{v_j \in V_i}(1 - p_j)\Bigr)$$

- its probabilistic version becomes a particular weighted version of the initial problem where a vertex $v_j \in V$ is assigned weight $1 - p_j$, a set V_i of the collection C

1. Any feasible solution of which is a subset V' of the vertex-set V of the input graph G.

2. Any feasible solution of which is a collection $C = (V_1, V_2, \ldots, V_k)$ of subsets of V verifying some non-trivial hereditary property.

is assigned with weight $w(V_i) = (1 - \prod_{v_j \in V_i}(1 - p_j))$ and the objective becomes to optimize the total weight of a solution (i.e., the quantity $\sum_{i=1}^{k} w(V_i)$);

- since setting $p_j = 1$, for any $v_j \in V$, $w(V_i) = 1$, the deterministic problem and its probabilistic counterpart coincide, the complexity of the latter is the one of the former.

– Finally, when mentioned that a problem[3] fits conditions of Theorem 6.3 in Chapter 6, then:

- its functional associated with an *a priori* solution E' is expressed as:

$$E(G, E') = \sum_{(v_i, v_j) \in E'} p_i p_j$$

- its probabilistic version becomes an edge-weighted version, where a weight equal to the product of the presence-probabilities of its endpoints is associated with any edge of E, and where the objective becomes to optimize the total weight of a solution, i.e., the sum of the weights of the edges in E';

- since setting $p_i = 1$, for any $v_i \in V$, the deterministic problem and its probabilistic counterpart coincide, the complexity of the latter is the one of the former.

7.1. Covering and partitioning

7.1.1. MIN VERTEX COVER

Consider a graph $G(V, E)$. A *vertex cover* of G is a subset $V' \subseteq V$ such that, for any $(v, u) \in E$, either $u \in V'$ or $v \in V'$. The objective for MIN VERTEX COVER is to determine a minimum-size vertex cover of G.

As we have seen in Chapter 3, MS is feasible for PROBABILISTIC MIN VERTEX COVER. Furthermore, the optimal *a priori* solution for this problem corresponds to the optimal solution of a particular weighted version where any vertex is weighted by its probability, and the objective is to determine a minimum-weight vertex cover. Consequently, PROBABILISTIC MIN VERTEX COVER is **NP**-hard (see also Chapter 6, Theorem 6.1).

7.1.2. MIN COLORING

Consider a graph $G(V, E)$. In MIN COLORING, we wish to color V with as few colors as possible, so that no two adjacent vertices receive the same color. The *chromatic number* of G, denoted by $\chi(G)$, is the smallest number of colors that can feasibly

---

3. Any feasible solution of which is a subset E' of the edge-set E of the input graph G.

color its vertices. A graph G is called *k-colorable* if its vertices can be legally colored by k colors, in other words, if its chromatic number is at most k; it will be called *k-chromatic*, if k is its chromatic number.

As we have in Chapter 5, strategy MS is feasible for PROBABILISTIC MIN COLORING. Under MS, the functional associated with an *a priori* solution $C = (S_1, S_2, \ldots)$ is a type of minimum-weight coloring where the weight of a color $S_i = \{v_{i_1}, v_{i_2}, \ldots\}$ is given by $f(S_i) = 1 - \prod_{v_{i_j} \in S_i}(1 - p_{i_j})$, and the weight of C is given by $\sum_{S_i \in C} f(S_i)$. This functional can be computed in polynomial time and the optimal *a priori* solution is a coloring that minimizes the above sum. Since if the probability of any vertex is 1, one recovers the classical MIN COLORING, PROBABILISTIC MIN COLORING is **NP**-hard (see also Theorem 6.2 in Chapter 6).

7.1.3. MAX ACHROMATIC NUMBER

In MAX ACHROMATIC NUMBER we wish determine the maximum-size partition of V into independent sets such that no union of two of them constitutes an independent set. In other words, we wish to color the vertices of G using a maximum set of colors in such a way, for two colors S_i and S_j, their union does not form an independent set. The maximum number of colors verifying this condition is called an *achromatic number* of G.

Strategy MS is not feasible for MAX ACHROMATIC NUMBER, since deletion of a vertex from one of the *a priori* colors may result in a coloring for which the union of two of its colors still remains an independent set.

7.1.4. MIN DOMINATING SET

Consider a graph $G(V, E)$. A *dominating set* is a subset $V' \subseteq V$ such that for any $v \in V \setminus V'$, there exists $u \in V'$ such that $(u, v) \in E$ (we say that V' dominates V). In MIN DOMINATING SET the objective is to determine a minimum-cardinality dominating set.

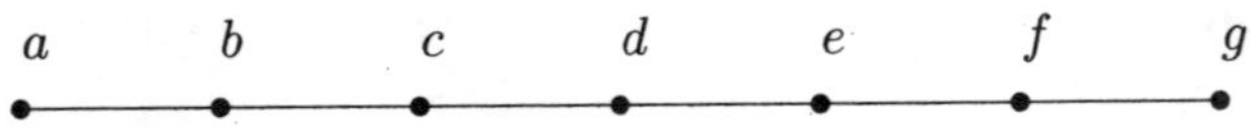

Figure 7.1. *MS is not always feasible for* MIN DOMINATING SET

Strategy MS is not always feasible, since removal of a vertex in V' may result in a set that does not dominate V any longer. Consider a path $\{a, b, c, d, e, f, g\}$. There, $V' = \{a, d, g\}$ dominates V, while if any of the vertices of V', say a, disappears, the resulting set V' is no more a dominating set, since vertex b becomes non-dominated (Figure 7.1).

7.1.5. MAX DOMATIC PARTITION

Here, the objective is to determine a maximum-size partition of the vertex-set V of a graph $G(V, E)$ into dominating sets.

As previously, in section 7.1.4, MS is not always feasible since, as noticed, removal of a vertex may result in a set that is no more dominating.

7.1.6. MIN EDGE-DOMINATING SET

Given a graph $G(V, E)$, an *edge-dominating set* is a subset $E' \subseteq E$ such that for any $e' \in E \setminus E'$, there exists $e \in E'$ such that e and e' are adjacent, i.e., they have a common endpoint. In MIN EDGE-DOMINATING SET, the objective is to determine a minimum-size edge-dominating set.

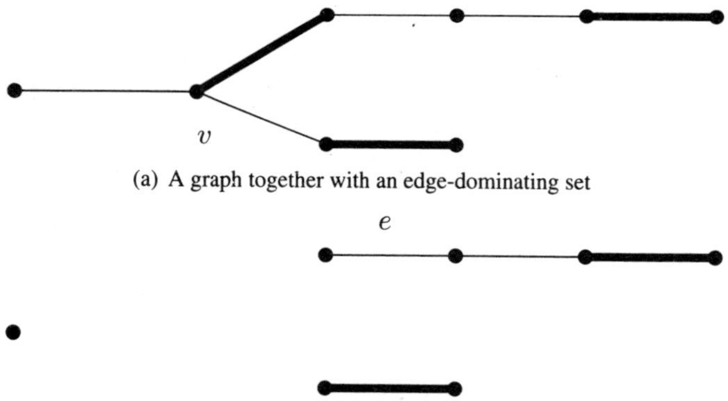

(a) A graph together with an edge-dominating set

(b) Heavy-lined edges are no more an edge-domintating set

Figure 7.2. *MS is not always feasible for* MIN EDGE-DOMINATING SET

Strategy MS is not feasible for MIN EDGE-DOMINATING SET. Figure 7.2 illustrates this fact. Indeed, consider the set of heavy-lined edges of Figure 7.2(a); this set is an edge-dominating set for the graph dealt. Removal of vertex v (together with the edges incident to it) results in a graph where the set of the (surviving) heavy-lined edges is no more edge-dominating (in Figure 7.2(b), edge e is not dominated).

7.1.7. MIN INDEPENDENT DOMINATING SET

Given a graph $G(V, E)$, we wish to determine a minimum-size independent set that is maximal for the inclusion; in other words, since a maximal independent set is always a dominating set, we search for a minimum-size maximal independent set.

For the same reasons as in section 7.1.4, MS is not always feasible for MIN INDEPENDENT DOMINATING SET.

7.1.8. MIN CHROMATIC SUM

Given a graph $G(V, E)$, we wish to compute a coloring $C = (S_1, \ldots S_k)$ of V that minimizes the quantity $\sum_{i=1}^{k} \sum_{v \in S_i} i$. This problem is **NP**-hard since it contains MIN COLORING as a subproblem[4].

Since the feasibility conditions imply that any of S_i is just an independent set, strategy MS is feasible for MIN CHROMATIC SUM. Note, however, that it does not fit Condition 4 of Theorem 6.2, which does not apply here. Let $V' \subseteq V$ be the set of the finally present vertices and denote by $1_{\{v_i \in V'\}}$ the indicator function of the fact: $v_i \in V'$. Then:

$$\begin{aligned} E(G, C) &= \sum_{V' \subseteq V} \Pr[V'] \sum_{i=1}^{k} \sum_{v \in S_i} i 1_{\{v \in V'\}} \\ &= \sum_{i=1}^{k} \sum_{v \in S_i} i \sum_{V' \subseteq V} \Pr[V'] \, 1_{\{v \in V'\}} \\ &= \sum_{i=1}^{k} i \left(\sum_{v_j \in S_i} p_j \right) \end{aligned} \qquad [7.1]$$

From [7.1], one can derive a tight characterization for the optimal *a priori* solution of PROBABILISTIC MIN CHROMATIC SUM. Indeed, if we consider that the weight $w(S_i)$ of a color S_i is the sum of the probabilities of its vertices multiplied by i, then the *a priori* solution C^* optimizing $E(G, C)$ is the coloring minimizing the quantity $\sum_{i=1}^{k} w(S_i)$. This is a type of weighted version of MIN CHROMATIC SUM where any vertex of the input-graph is weighted by its probability. This version being a generalization of the original MIN CHROMATIC SUM, it is also **NP**-hard.

4. In MIN COLORING, the weight of color S_i is 1 and not k_i.

7.1.9. MIN EDGE COLORING

Given a graph $G(V, E)$, the objective for MIN EDGE COLORING is to determine a minimum-size partition $D = (E_1, \ldots, E_k)$ of E such that, for all $i \in \{1, \ldots, k\}$, no two edges in E_i share some common endpoint. In other words, any of the E_is is a matching of G and MIN EDGE COLORING consists exactly of determining a *minimum-size partition into matchings*.

Strategy MS is feasible for MIN EDGE COLORING. Indeed, if a vertex disappears, then all of its incident edges disappear also. This removal does not violate the matching property for any of the subsets E_i of any *a priori* solution D. We are now going to compute $E(G, D)$. Denoting for $i = 1, \ldots, k$ by $1_{\{E_i \neq \emptyset\}}$ the indicator function of the fact: $E_i \neq \emptyset$, and by E' the edge-set of $G[V']$, we get:

$$
\begin{aligned}
m(G, S(V')) &= \sum_{i=1}^{k} 1_{\{E_i \neq \emptyset\}} = \sum_{i=1}^{k} 1_{\{E_i \cap E' \neq \emptyset\}} \\
&= \sum_{i=1}^{k} 1 - \sum_{i=1}^{k} 1_{\{E_i \cap E' = \emptyset\}}
\end{aligned}
\qquad [7.2]
$$

Note that $E_i \cap E' = \emptyset$ if and only if all the edges of E_i are absent. Note also that all the edges of E_i are independent, i.e., they do not share common vertices. So, assuming that $E_i = \{e_{i_1}, e_{i_2}, \ldots, e_{i_{j_i}}\}$ and taking into account that $\sum_{V' \subseteq V} \Pr[V'] = 1$, [7.2] becomes:

$$
m(G, S(V')) = \sum_{i=1}^{k} 1 - \sum_{i=1}^{k} 1_{\{E' \cap \{e_{i_1}\} = \emptyset\}} \times \ldots \times 1_{\left\{E' \cap \left\{e_{i_{j_i}}\right\} = \emptyset\right\}} \qquad [7.3]
$$

Using [7.3], the functional $E(G, D)$ becomes:

$$
\begin{aligned}
E(G, D) &= \sum_{V' \subseteq V} \Pr[V'] m(G, S(V')) \\
&= \sum_{V' \subseteq V} \Pr[V'] \left(\sum_{i=1}^{k} 1 - \sum_{i=1}^{k} 1_{\{E' \cap \{e_{i_1}\} = \emptyset\}} \times \ldots \times 1_{\left\{E' \cap \left\{e_{i_{j_i}}\right\} = \emptyset\right\}} \right) \\
&= \sum_{i=1}^{k} 1 - \sum_{i=1}^{k} \sum_{V' \subseteq V} \Pr[V'] 1_{\{E' \cap \{e_{i_1}\} = \emptyset\}} \times \ldots \times 1_{\left\{E' \cap \left\{e_{i_{j_i}}\right\} = \emptyset\right\}}
\end{aligned}
\qquad [7.4]
$$

Observe now that, setting $e = (v_k, v_l)$, we have:

$$1_{\{E' \cap \{e\} = \emptyset\}} = 1 - 1_{\{E' \cap \{e\} \neq \emptyset\}} = 1 - 1_{\{V' \cap \{v_k\} \neq \emptyset\}} \times 1_{\{V' \cap \{v_l\} \neq \emptyset\}} \qquad [7.5]$$

the above equality meaning that in order that an edge is present in E', both of these endpoints have to be present in V'.

From [7.4] and [7.5], we immediately get:

$$E(G, D) = k - \sum_{i=1}^{k} \prod_{(v_k, v_l) \in E_i} (1 - p_k p_l) = \sum_{i=1}^{k} \left(1 - \prod_{(v_k, v_l) \in E_i} (1 - p_k p_l) \right)$$

From the above expression for the functional, one derives that considering weight $w(v_i, v_j) = 1 - p_i p_j$, for any edge $(v_i, v_j) \in E$ and considering as value of a matching E_k of G the quantity:

$$v\left(E_k\right) = 1 - \prod_{(v_i, v_j) \in E_k} (1 - p_i p_j) = 1 - \prod_{(v_i, v_j) \in E_k} w\left(v_i, v_j\right)$$

the optimal *a priori* solution for PROBABILISTIC MIN EDGE COLORING is a partition $D = (E_1, E_2, \ldots)$ of E into matchings minimizing the quantity $\sum_{E_i \in D} v(E_i)$. Since, for $p_i = 1$, $v_i \in V$, one recovers the standard MIN EDGE COLORING, one immediately deduces that PROBABILISTIC MIN EDGE COLORING is **NP**-hard.

7.1.10. MIN FEEDBACK VERTEX-SET

Given an oriented graph $G(V, A)$, a *feedback vertex-set* is a subset $V' \subseteq V$ such that V' contains at least a vertex of any directed cycle of G. In MIN FEEDBACK VERTEX-SET, the objective is to determine a feedback vertex-set of minimum size.

Note that absence of a vertex v from V' breaks any cycle containing this vertex. If v makes part of an *a priori* solution S then, since no such cycle that contained v exists in G', feasibility of the solution returned by `MS` does not suffer from the absence of v. So, `MS` is feasible for MIN FEEDBACK VERTEX-SET. and Theorem 6.1 in Chapter 6 applies for this problem. Hence, PROBABILISTIC MIN FEEDBACK VERTEX-SET is **NP**-hard.

7.1.11. MIN FEEDBACK ARC-SET

Given an oriented graph $G(V, A)$, a *feedback edge-set* is a subset $A' \subseteq A$ such that A' contains at least an arc of any directed cycle of G. In MIN FEEDBACK ARC-SET, the objective is to determine a feedback arc-set of minimum size.

With exactly similar arguments as for the case of MIN FEEDBACK VERTEX-SET (section 7.1.10), one can conclude that MS is feasible for MIN FEEDBACK ARC-SET also. Hence, Theorem 6.3 of section 6.1.3 in Chapter 6 applies for this problem. Consequently, PROBABILISTIC MIN FEEDBACK ARC-SET is **NP**-hard.

7.1.12. MAX MATCHING

In MAX MATCHING, the objective is to determine a maximum-size matching (see section A.2 in Appendix A). This problem has been studied in section 6.1.1 and shown to fit conditions of Theorem 6.1. Consequently, PROBABILISTIC MAX MATCHING is in **P**.

7.1.13. MIN MAXIMAL MATCHING

In MIN MAXIMAL MATCHING, the objective is to determine a minimum-size maximal matching (see section A.2 in Appendix A). It is easy to see that the maximality constraint makes that strategy MS is not always feasible for the problem dealt. Indeed, absence of a vertex in G' (the finally present subgraph) might entail the absence of an edge from the *a priori* solution S in such a way that the resulting matching might not be maximal.

7.1.14. MAX TRIANGLE PACKING

Given a graph $G(V, E)$, a vertex-packing into triangles is a collection $(V_1, \ldots, V_k)$ such that, for any $i = 1, \ldots, k$, $V_i \subseteq V$, $|V_i| = 3$ and the subgraph induced by V_i is a triangle (i.e., a complete graph on 3 vertices) and, for any $i, j = 1, \ldots, k$, $V_i \cap V_j = \emptyset$. In MAX TRIANGLE PACKING, the objective is to determine a maximum such collection.

Modification strategy MS is not feasible for the problem dealt since absence of a vertex from any *a priori* solution S possibly breaks a triangle; hence the surviving solution is not a packing into triangles. However, if one slightly modifies MS in such a way that, once a triangle broken, it is removed from the solution, the rest of this solution constitutes a vertex packing. The strategy thus modified, the functional of which is studied in the sequel, then becomes feasible for MAX TRIANGLE PACKING. Furthermore, note that this problem cannot be handled as an application of Theorem 6.2 in Chapter 6, because property "is a triangle" is not hereditary (Condition 2).

Setting $(V_1, V_2, \ldots, V_k)$ and, for $i = 1, \ldots, k$, $V_i = \{v_i, u_i, w_i\}$, we get:

$$\begin{aligned} E(G,S) &= \sum_{V' \subseteq V} \Pr[V'] m(G, S(V')) \\ &= \sum_{V' \subseteq V} \Pr[V'] \sum_{i=1}^{k} 1_{\{v_i, u_i, w_i \in V'\}} \\ &= \sum_{i=1}^{k} \sum_{V' \subseteq V} \Pr[V'] 1_{\{v_i, u_i, w_i \in V'\}} = \sum_{i=1}^{k} p_{v_i} p_{u_i} p_{w_i} \end{aligned} \quad [7.6]$$

From [7.6], we immediately conclude that the optimal *a priori* solution for PROBABILISTIC MAX TRIANGLE PACKING is the optimal solution of a weighted version of MAX TRIANGLE PACKING, where any vertex of V is weighted by its presence-probability. Hence, PROBABILISTIC MAX TRIANGLE PACKING is **NP**-hard.

7.1.15. MAX H-MATCHING

Consider a graph $G(V, E)$ and a connected subgraph H of G of order at least 3. An H-matching of G is a collection $G_1, G_2, \ldots$ of vertex-disjoint subgraphs of G such that any subgraph G_i in this collection is isomorphic[5] to H. In MAX H-MATCHING, the objective is to determine a maximum size H-matching.

With a reasoning exactly similar to the one of section 7.1.14, and setting:

$$S = (G_1, G_2, \ldots, G_k)$$

an *a priori* solution for MAX H-MATCHING, $V_1, \ldots, V_k$ the vertex-sets of $G_1, \ldots, G_k$, respectively, and, for $i = 1, \ldots, k$, $V_i = \{v_{i_1}, \ldots, v_{i_{|V_i|}}\}$, we get:

$$E(G,S) = \sum_{i=1}^{k} \prod_{j=1}^{|V_i|} p_{v_{i_j}} \quad [7.7]$$

---

5. Two graphs $G_1(V_1, E_1)$ and $G_2(V_2, E_2)$ are said to be *isomorphic* if there exist two one-to-one functions $f : V_1 \to V_2$ and $g : E_1 \to E_2$ such that, for any edge $(u, v) \in E_1$, edge $(f(u), f(v))$ is via g in E_2.

From [7.7], we immediately conclude that the optimal *a priori* solution for PROBABILISTIC MAX H-MATCHING is the optimal solution of a weighted version of MAX H-MATCHING, where any vertex of V is weighted by its presence-probability. Hence, PROBABILISTIC MAX H-MATCHING is **NP**-hard.

7.1.16. MIN PARTITION INTO CLIQUES

Given a graph $G(V, E)$, MIN PARTITION INTO CLIQUES consists of determining a minimum size partition $(V_1, V_2, \ldots)$ of V such that the subgraph induced by any of the V_is is a clique of G. It is easy to see that any coloring of G induces a partition into cliques of the same size in the complement $\bar{G}$ (see section A.2 of Appendix A) of G.

It is easy to see that, given an *a priori* solution $S = (V_1, V_2, \ldots, V_k)$ for MIN PARTITION INTO CLIQUES, if a vertex belonging to say V_i for some $i \leqslant k$ disappears, the rest of the vertices of V_i always induces a clique; hence, MS is feasible for the problem covered and, furthermore, property "is a clique" is hereditary. So, MIN PARTITION INTO CLIQUES meets the conditions of Theorem 6.2 in Chapter 6; so, PROBABILISTIC MIN PARTITION INTO CLIQUES is **NP**-hard.

7.1.17. MIN CLIQUE COVER

With the same arguments as in section 7.1.16, one can derive that this problem also fits perfectly the conditions of Theorem 6.2 in Chapter 6. So, the results stated for MIN PARTITION INTO CLIQUES in section 7.1.16 hold identically for PROBABILISTIC MIN CLIQUE COVER.

7.1.18. MIN k-CAPACITED TREE PARTITION

Given a graph $G(V, E)$; an edge-weight function $w : E \rightarrow \mathbb{N}$ and an integer k, a partition into k-capacited trees is a collection $(E_1, E_2, \ldots, E_m)$ of pairwise disjoint subsets of E such that $\cup_{i=1}^{m} E_i = E$ and the subgraph induced by any of the E_is is a tree of order at least k. In MIN k-CAPACITED TREE PARTITION, the objective is to determine such a partition minimizing its total weight, i.e., the quantity $\sum_{i=1}^{m} \sum_{e \in E_i} w(e)$.

Strategy MS is not feasible for MIN k-CAPACITED TREE PARTITION since absence of a vertex entails absence of some edges, the removal of which might disconnect one of more of the trees of the partition, the new trees thus produced having order less than k.

7.1.19. MAX BALANCED CONNECTED PARTITION

Consider a connected graph $G(V, E)$ and vertex-weight function $w : V \rightarrow \mathbb{N}$. A connected partition (V_1, V_2) of V is a pair of disjoint subsets V_1 and V_2 of V such that $V_1 \cup V_2 = V$ and the subgraphs of G induced by both V_1 and V_2 are connected. MAX BALANCED CONNECTED PARTITION consists of determining a connected partition (V_1, V_2) maximizing its balance, i.e., the quantity $\min\{w(V_1), w(V_2)\}$, where for a subset $V' \in V$, $w(V') = \sum_{v \in V'} w(V')$.

Strategy MS is not feasible for this problem. Indeed, absence of a vertex from one of V_1, or V_2 may disconnect the corresponding subgraph.

7.1.20. MIN COMPLETE BIPARTITE SUBGRAPH COVER

Given a graph $G(V, E)$, a feasible solution of MIN COMPLETE BIPARTITE SUBGRAPH COVER is a collection $\mathcal{C} = (V_1, V_2, \dots, V_k)$ of subsets of V such that the subgraph induced by any of the V_is, $i = 1, \dots, k$, is a complete bipartite graph and for any edge $(u, v) \in E$ there exists a V_i containing both u and v. The objective here is to minimize the size $|\mathcal{C}|$ of $\mathcal{C}$.

Modification strategy MS is feasible for MIN COMPLETE BIPARTITE SUBGRAPH COVER. Indeed, if a vertex v disappears from some subset V_i of an *a priori* solution $\mathcal{C}$, the surviving set V_i always induces a complete bipartite graph; furthermore:

– except for the edges that have been disappeared (the ones incident to v), any other edge remain covered by the surviving sets of $\mathcal{C}$;

– property "is a complete bipartite graph" is hereditary.

Since MS is feasible for MIN COMPLETE BIPARTITE SUBGRAPH COVER, this problem fits the conditions of Theorem 6.2 in Chapter 6. Consequently, PROBABILISTIC MIN COMPLETE BIPARTITE SUBGRAPH COVER is **NP**-hard.

7.1.21. MIN VERTEX-DISJOINT CYCLE COVER

Given a graph $G(V, E)$, the objective in MIN VERTEX-DISJOINT CYCLE COVER is to cover the vertices of V by a collection $(V_1, \dots, V_k)$ of pair-wise disjoint subsets of V such that any of the V_is induces a cycle as partial subgraph.

Strategy MS is not feasible for MIN VERTEX-DISJOINT CYCLE COVER, since absence of a vertex from the *a priori* solution might break one of its cycles.

7.1.22. MIN CUT COVER

Given a graph $G(V, E)$, a feasible solution for MIN CUT COVER is a collection $(V_1, \ldots, V_k)$ of V such that any V_i, $i = 1, \ldots, k$ is a cut, i.e., for any $(u, v) \in E$, there exists a V_i such that either $u \in V_i$ and $v \notin V_i$, or $u \notin V_i$ and $v \in V_i$. The objective is to minimize the size of the collection.

Consider an *a priori* solution $S = (V_1, \ldots, V_k)$. If a vertex $v \in V$ is absent, then any edge incident to v is also absent. However, absence of a vertex together with any edge incident to it does not affect the edges present to the final graph $G'(V', E')$, which remain feasibly covered by endpoints, any of them belonging to distinct cuts. Hence, strategy MS is feasible for MIN CUT COVER, which meets the conditions of Theorem 6.2 in Chapter 6, since property "is a cut" is hereditary. Hence, PROBABILISTIC MIN CUT COVER is **NP**-hard.

7.2. Subgraphs and supergraphs

7.2.1. MAX INDEPENDENT SET

Consider a graph $G(V, E)$. An *independent set* is a subset $V' \subseteq V$ such that, for any $(v, u) \in V' \times V'$, $(u, v) \notin E$. The objective for MAX INDEPENDENT SET is to determine an independent set of maximum size in G.

As we have seen in Chapter 2, MS is feasible for PROBABILISTIC MAX INDEPENDENT SET. Furthermore, the optimal *a priori* solution for this problem corresponds to the optimal solution of particular weighted version where any vertex is weighted by its probability, and the objective is to determine a maximum-weight independent set. Consequently, PROBABILISTIC MAX INDEPENDENT SET is **NP**-hard (see also Chapter 6, Theorem 6.1).

7.2.2. MAX CLIQUE

Consider a graph $G(V, E)$. A *clique* of G is a complete subgraph of G induced by a subset $V' \subseteq V$. The objective for MAX CLIQUE is to determine a maximum-size subset $V' \subset V$ such that the subgraph of G induced by V' is a clique.

Strategy MS is obviously feasible for MAX CLIQUE (see also section A.2 in Appendix A). So, this problem fits the conditions of Theorem 6.1 in Chapter 6. Consequently, PROBABILISTIC MAX CLIQUE is **NP**-hard.

7.2.3. MAX INDEPENDENT SEQUENCE

Consider a graph $G(V, E)$. An *independent sequence* of G is a sequence of independent vertices of G such that, for all i, a vertex v_i exists which is adjacent to v_{i+1} but is not adjacent to any v_j for $j < i$. In MAX INDEPENDENT SEQUENCE, the objective is to find a maximum-length independent sequence.

Strategy MS is not feasible for this problem. Indeed, if a vertex, say v_{i+1} of an *a priori* (feasible) sequence S is absent, then, since S is feasible, vertex v_{i+2} will not be adjacent to v_i. Consequently, the resulting subsequence will be unfeasible.

7.2.4. MAX INDUCED SUBGRAPH WITH PROPERTY π

Consider a graph $G(V, E)$ and a non-trivial hereditary property π. A feasible solution for MAX INDUCED SUBGRAPH WITH PROPERTY π is a subset $V' \subseteq V$ such that the subgraph $G[V']$ of G induced by V' satisfies π. The objective for MAX INDUCED SUBGRAPH WITH PROPERTY π is to determine such a set V' of maximum-size. Note that "*independent set*", "*clique*", "*planar graph*" are hereditary properties.

Strategy MS is feasible for MAX INDUCED SUBGRAPH WITH PROPERTY π, since, by the definition of π, if a subset $S \subseteq V$ (an *a priori* solution) induces a subgraph verifying it, then any subset of S (resulting from removal of any of its vertices) also induces a subgraph verifying π. Henceforth, MAX INDUCED SUBGRAPH WITH PROPERTY π fits the conditions of Theorem 6.1 in Chapter 6; consequently, PROBABILISTIC MAX INDUCED SUBGRAPH WITH PROPERTY π is **NP**-hard.

7.2.5. MIN VERTEX DELETION TO OBTAIN SUBGRAPH WITH PROPERTY π

Consider a graph $G(V, E)$ and a non-trivial hereditary property π. A feasible solution of MIN VERTEX DELETION TO OBTAIN SUBGRAPH WITH PROPERTY π is a subset $V' \subseteq V$ such that the subgraph $G[V \setminus V']$ induced by $V \setminus V'$ verifies π; the objective is to determine a minimum-cardinality set V' to be removed.

Given an *a priori* solution S, i.e., a set of vertices such that $G[V \setminus S]$ verifies π, the absence of a vertex v from S has no impact to the set $(V \setminus \{v\}) \setminus (S \setminus \{v\}) = V \setminus S$ that always verifies π. So, MS is feasible for MIN VERTEX DELETION TO OBTAIN SUBGRAPH WITH PROPERTY π, that verifies all of the conditions of Theorem 6.1 in Chapter 6. Consequently, PROBABILISTIC MIN VERTEX DELETION TO OBTAIN SUBGRAPH WITH PROPERTY π is **NP**-hard.

7.2.6. MIN EDGE DELETION TO OBTAIN SUBGRAPH WITH PROPERTY π

Given a graph $G(V, E)$ and a non-trivial hereditary property π, MIN EDGE DELETION TO OBTAIN SUBGRAPH WITH PROPERTY π consists of determining a minimum size set $E' \subseteq E$, such that the partial subgraph $G' = G(V, E \setminus E')$ verifies π.

Given an *a priori* solution $E' \subseteq E$, the absence of a vertex v will entail removal of the edges incident to it, some of them being in E'. In any case, the absence of a set X of vertices from G will produce, with respect to G' an induced subgraph $G'[V \setminus X]$ that, being a subgraph of G', will verify π (which is already verified by G' itself). Consequently, MS is feasible for MIN EDGE DELETION TO OBTAIN SUBGRAPH WITH PROPERTY π that fits all the conditions of Theorem 6.3 in Chapter 6. Consequently, PROBABILISTIC MIN EDGE DELETION TO OBTAIN SUBGRAPH WITH PROPERTY π is **NP**-hard.

7.2.7. MAX CONNECTED SUBGRAPH WITH PROPERTY π

Given a graph $G(V, E)$ and a non-trivial hereditary property π, MAX CONNECTED SUBGRAPH WITH PROPERTY π consists of determining a maximum-size subset $V' \subseteq V$, such that $G[V']$ is connected and verifies π.

It is easy to see that MS is not feasible for MAX CONNECTED SUBGRAPH WITH PROPERTY π, since removal of some vertices from V' may disconnect the graph induced by this subset.

7.2.8. MIN VERTEX DELETION TO OBTAIN CONNECTED SUBGRAPH WITH PROPERTY π

Given a graph $G(V, E)$ and a non-trivial hereditary property π, MIN VERTEX DELETION TO OBTAIN CONNECTED SUBGRAPH WITH PROPERTY π consists of determining a minimum-size subset $V' \subseteq V$, such that $G[V \setminus V']$ is connected and verifies π.

Strategy MS is not feasible for the same reasons as in section 7.2.7.

7.2.9. MAX DEGREE-BOUNDED CONNECTED SUBGRAPH

Given a graph $G(V, E)$, a weight function $w : E \rightarrow \mathbb{N}$, and an integer $d \geqslant 2$, a feasible solution for MAX DEGREE-BOUNDED CONNECTED SUBGRAPH is a subset $E' \subseteq E$, such that the partial subgraph $G(V, E')$ of G is connected and has degree bounded by d. The objective for MAX DEGREE-BOUNDED CONNECTED SUBGRAPH is to maximize the total weight of E', i.e., the quantity $\sum_{e \in E'} w(e)$.

Strategy MS is not feasible for the same reasons as in sections 7.2.7 and 7.2.8.

7.2.10. MAX PLANAR SUBGRAPH

Given a graph $G(V, E)$, MAX PLANAR SUBGRAPH consists of determining a maximum-size subset $E' \subseteq E$, such that the partial subgraph $G'(V, E')$ of G is planar.

Absence of a vertex from V may entail removal of some edges in E' without violating the planarity property of the subgraph of G' induced by the surviving vertices. Hence, MS is feasible for MAX PLANAR SUBGRAPH which fits conditions of Theorem 6.3 in Chapter 6. Consequently, PROBABILISTIC MAX PLANAR SUBGRAPH is **NP**-hard. Note also that this problem is a particular restriction of PROBABILISTIC MIN EDGE DELETION TO OBTAIN SUBGRAPH WITH PROPERTY π, as seen in section 7.2.6.

7.2.11. MIN EDGE DELETION k-PARTITION

Consider a graph $G(V, E)$ together with an edge-weight function $w : E \rightarrow \mathbb{N}$. A feasible solution for MIN EDGE DELETION k-PARTITION is a k-partition, i.e., some color assignment $c : V \rightarrow \{1, 2, \ldots, k\}$ and the objective is to minimize the total weight of the monochromatic edges, i.e., the quantity:

$$\sum_{\substack{(v_i, v_j) \in E \\ c(v_i) = c(v_j)}} w\left(v_i, v_j\right)$$

In other words, the objective of MIN EDGE DELETION k-PARTITION is to determine the minimum total-weight edge-subset, the removal of which transforms G into a k-partite graph[6].

Obviously, given an *a priori* coloring C, absence of some vertices from V does not change the coloring of the remaining vertices; so MS is feasible for MIN EDGE DELETION k-PARTITION. Note now that the latter of the two alternative definitions of this problem presented just above, together with the feasibility of MS for it, make MIN EDGE DELETION k-PARTITION perfectly fit the conditions of Theorem 6.3 in Chapter 6. Consequently, PROBABILISTIC MIN EDGE DELETION k-PARTITION is **NP**-hard.

7.2.12. MAX k-COLORABLE SUBGRAPH

A graph G is called *k-colorable* if its vertices can be colored with at most k colors, i.e., if its chromatic number is less than, or equal to, k. Consider a graph $G(V, E)$. In

6. A graph G is said to be k-partite if its vertices can be partitioned into a collection $(V_1, \ldots, V_k)$ of independent sets; in a bipartite graph, $k = 2$.

MAX k-COLORABLE SUBGRAPH the objective is to determine a maximum-size subset $E' \subseteq E$ such that the partial subgraph $G'(V, E')$ is k-colorable.

The absence of a vertex from V preserves the fact that the surviving subgraph of G' remains k-colorable. So, MS is feasible for MAX k-COLORABLE SUBGRAPH which fits conditions of Theorem 6.3 in Chapter 6. Hence, PROBABILISTIC MAX k-COLORABLE SUBGRAPH is **NP**-hard.

7.2.13. MAX SUBFOREST

Consider a graph $G(V, E)$ and a forest (i.e., a set of trees) H. In MAX SUBFOREST, the objective is to determine a maximum-size set $E' \subseteq E$, such that the partial subgraph $G'(V, E')$ of G does not contain any subtree isomorphic to a tree from H.

Strategy MS is not feasible for MAX SUBFOREST. Indeed, given an *a priori* solution for this problem, the absence of some vertices may transform a tree non-isomorphic to some tree of H into one that this time is isomorphic to one of the trees of H.

7.2.14. MAX EDGE SUBGRAPH *or* DENSE k-SUBGRAPH

Consider a graph $G(V, E)$, an edge-weight function $w : E \rightarrow \mathbb{N}$ and an integer k. In MAX EDGE SUBGRAPH, the objective is to determine a subset $V' \subseteq V$ with $|V'| = k$ such that the quantity $\sum_{(u,v) \in E \cap V' \times V'} w(u, v)$ is maximized.

Strategy MS is not feasible for MAX EDGE SUBGRAPH. Indeed, given an *a priori* solution V' of size k, the removal of a vertex from V' will produce a set of size smaller than k, thus violating the feasibility condition of the problem.

7.2.15. MIN EDGE K-SPANNER

Consider a connected graph $G(V, E)$. A *k-spanner* of G is a spanning subgraph G' of G such that, for any pair of vertices u and v, the length of the shortest path between u and v in G' is at most k times the distance between u and v in G. In MIN EDGE k-SPANNER, the objective is to determine a k-spanner of G with the smallest number of edges.

Obviously, strategy MS is not feasible for MIN EDGE k-SPANNER since the absence of some vertices can disconnect an *a priori* k-spanner G'.

7.2.16. MAX k-COLORABLE INDUCED SUBGRAPH

Given a graph $G(V, E)$, MAX k-COLORABLE INDUCED SUBGRAPH consists of determining a maximum-size subset $V' \subseteq V$, such that the subgraph $G[V']$ of G induced by V' is k-colorable.

Note first that "k-colorability" is a hereditary property. Consequently (see section 7.2.4), strategy MS is feasible for MAX k-COLORABLE INDUCED SUBGRAPH, which fits now conditions of Theorem 6.1. Hence, PROBABILISTIC MAX k-COLORABLE INDUCED SUBGRAPH is **NP**-hard.

7.2.17. MIN EQUIVALENT DIGRAPH

Consider a directed graph $G(V, A)$. In MIN EQUIVALENT DIGRAPH, the objective is to determine a minimum-cardinality subset $E' \subseteq E$ such that, for any ordered pair of vertices u and v in V, the partial graph $G'(V, E')$ contains a directed path from u to v if and only if G does.

Strategy MS is feasible for MIN EQUIVALENT DIGRAPH. Simply note that if a vertex disappears from G, then any path associated with this vertex (having it as initial endpoint) disappears also, the rest of the paths remaining unchanged in the surviving part of G'. So, PROBABILISTIC MIN EQUIVALENT DIGRAPH fits the conditions of Theorem 6.3 and is **NP**-hard.

7.2.18. MIN CHORDAL GRAPH COMPLETION

Consider a graph $G(V, E)$. MIN CHORDAL GRAPH COMPLETION consists of determining a minimum size superset $E' \supset E$, such that the graph $G'(V, E')$ is chordal (see section A.2 in Appendix A).

Strategy MS is feasible for MIN CHORDAL GRAPH COMPLETION (if a vertex disappears, the cycles to which it is contained are broken). So, PROBABILISTIC MIN CHORDAL GRAPH COMPLETION fits the conditions of Theorem 6.3 and is **NP**-hard.

7.3. Iso- and other morphisms

7.3.1. MAX COMMON SUBGRAPH

Consider two graphs $G_1(V_1, E_1)$ and $G_2(V_2, E_2)$. In MAX COMMON SUBGRAPH, the objective is to determine maximum-size subsets $E_1' \subseteq E_1$ and $E_2' \subseteq E_2$, such that the partial subgraphs $G_1'(V_1, E_1')$ and $G_2'(V_2, E_2')$ are isomorphic.

Since G_1' and G_2' are required to be isomorphic, we can assume that $|V_1| = |V_2| = |V|$ and that $|E_1'| = |E_2'| = |E'|$. Moreover, we assume that a vertex v disappears simultaneously from both V_1 and V_2. Under this last assumption, MS is feasible (it would not be the case otherwise). Indeed, simultaneous removal of a vertex from two isomorphic graphs leaves them isomorphic. Hence, MAX COMMON SUBGRAPH fits the conditions of Theorem 6.3 in Chapter 6 and, consequently, PROBABILISTIC MAX COMMON SUBGRAPH is **NP**-hard.

7.3.2. MAX COMMON INDUCED SUBGRAPH

Consider two graphs $G_1(V_1, E_1)$ and $G_2(V_2, E_2)$. MAX COMMON INDUCED SUBGRAPH consists of determining subsets $V_1' \subseteq V_1$ and $V_2' \subseteq V_2$ such that the subgraphs $G_1[V_1']$ and $G_2[V_2']$ of G_1 and G_2, induced by V_1' and V_2', respectively, are isomorphic.

For the same reasons as in section 7.3.1 for MAX COMMON SUBGRAPH, we assume that $|V_1'| = |V_2'| = |V'|$ and, furthermore, that a vertex v disappears simultaneously from both V_1 and V_2. In this case, as previously, MS is feasible and MAX COMMON INDUCED SUBGRAPH fits the conditions of Theorem 6.1 in Chapter 6. Consequently, PROBABILISTIC MAX COMMON INDUCED SUBGRAPH is **NP**-hard.

7.3.3. MAX COMMON EMBEDDED SUBTREE

Consider two node-labelled trees T_1 and T_2. A common embedded subtree is a labelled tree T' that can be embedded into both T_1 and T_2, an embedding from T' to T being an injective function from the nodes of T' to the ones of T that preserves labels and ancestorship[7]. In MAX COMMON EMBEDDED SUBTREE, the objective is to determine a maximum-order common embedded subtree.

Strategy MS is not feasible since the absence of a vertex from an *a priori* common embedded subtree might disconnect it.

7.3.4. MIN GRAPH TRANSFORMATION

Consider two graphs $G_1(V, E_1)$ and $G_2(V, E_2)$. In MIN GRAPH TRANSFORMATION, the objective is to determine a minimum-size subset E' of E_1 such that graphs $G_1'(V, E_1 \setminus E')$ and $G_2'(V, E_2 \cup E')$ become isomorphic.

7. Note that since fathership does not need to be preserved, T' does not need to be an ordinary subtree.

As in sections 7.3.1 and 7.3.2, we need to assume that vertices disappear simultaneously from both G'_1 and G'_2. Under this assumption, MS is feasible since the surviving subgraphs of both G'_1 and G'_2 remain isomorphic. So, MIN GRAPH TRANSFORMATION satisfies the conditions of Theorem 6.3 in Chapter 6 and, henceforth, PROBABILISTIC MIN GRAPH TRANSFORMATION is **NP**-hard.

7.4. Cuts and connectivity

7.4.1. MAX CUT

Consider a graph $G(V, E)$. In MAX CUT, we wish to determine a maximum cut that is to partition V into two subsets V_1 and V_2 such that a maximum number of edges have one of their endpoints in V_1 and the other one in V_2.

As we have seen in section 6.1.3 of Chapter 6, PROBABILISTIC MAX CUT fits the conditions of Theorem 6.3. Consequently its optimal *a priori* solution is the optimal solution of a particular edge-weighted MAX CUT-problem, where any edge (v_i, v_j) of the input-graph is weighted by $p_i p_j$. Based on this characterization, PROBABILISTIC MAX CUT is **NP**-hard.

7.4.2. MAX DIRECTED CUT

Consider a directed graph $G(V, E)$. In MAX DIRECTED CUT, we wish to determine a maximum directed cut, i.e., to partition V into two subsets V_1 and V_2 such that a maximum number of arcs have one of their endpoints, say the initial, in V_1 and their terminal endpoint in V_2.

For this problem, hold exactly the same results as for its undirected version of section 7.4.1.

7.4.3. MIN CROSSING NUMBER

Consider a directed graph $G(V, A)$. In MIN CROSSING NUMBER, the objective is to compute an embedding of G in the plane, minimizing the number of pairs of edges crossing one another.

Strategy MS is feasible for MIN CROSSING NUMBER since, even if some vertices disappear, the embedding of the surviving subgraph of G in the plane remains feasible.

Since this problem does not belong to any of the cases treated by the three main theorems of Chapter 6, we are going to compute explicitly its functional associated with MS. Consider an *a priori* embedding R of G in the plane. Then, denoting by $c_1, \ldots, c_r$, the crossings in R, we get:

$$E(G,R) = \sum_{V' \subseteq V} \Pr[V'] \, m(G, R(V'))$$
$$= \sum_{V' \subseteq V} \Pr[V'] \sum_{i=1}^{r} 1_{\{c_i \in R(V')\}} \quad [7.8]$$

Consider crossing c_m due to edges (v_{m_i}, v_{m_j}) and (v_{m_k}, v_{m_l}). Observe that c_m survives in $R(V')$ if and only if all the four vertices v_{m_i}, v_{m_j}, v_{m_k} and v_{m_l} are present in the subgraph of G induced by the finally present subset V' of V, i.e., if both of the edges (v_{m_i}, v_{m_j}) and (v_{m_k}, v_{m_l}) are present. Based upon this observation, [7.8] becomes:

$$E(G,R) = \sum_{i=1}^{r} \sum_{V' \subseteq V} \Pr[V'] \, 1_{\{c_i \in R(V')\}} = \sum_{i=1}^{r} p_{i_i} p_{i_j} p_{i_k} p_{i_l} \quad [7.9]$$

From [7.9], one derives that an optimal *a priori* solution for PROBABILISTIC MIN CROSSING NUMBER is an optimal solution of a particular weighted version of MIN CROSSING NUMBER where any edge is weighted by the product of the presence-probabilities of its endpoints, a crossing is weighted by the product of the weights of the edges generating it, and the objective is to compute an embedding minimizing the total weight of the crossings. Obviously, setting $p_i = 1$, for any $v_i \in V$, we recover the classical MIN CROSSING NUMBER. So, PROBABILISTIC MIN CROSSING NUMBER is **NP**-hard.

7.4.4. MAX k-CUT

Consider a graph $G(V, E)$ of order n, an edge-weight function $w : E \rightarrow \mathbb{N}$, and an integer $k \in \{2, \ldots, n\}$. In MAX k-CUT, the objective is to determine a partition of V into k disjoint sets $V_1, V_2, \ldots, V_k$ such that the quantity:

$$\sum_{i=1}^{k-1} \sum_{j=i+1}^{k} \sum_{\substack{v_l \in V_i \\ v_m \in V_j \\ (v_l, v_m) \in E}} w(v_l, v_m) \quad [7.10]$$

is maximized.

Considering an *a priori* k-cut, absence of some vertices may result in a cut implying less than k subsets of V. Hence, MS is not always feasible for MAX k-CUT.

7.4.5. MIN k-CUT

This is the same problem as MAX k-CUT (section 7.4.4) up to the fact that the goal now is to minimize [7.10]. The conclusions are the same as were given previously in section 7.4.4.

7.4.6. MIN NETWORK INHIBITION ON PLANAR GRAPHS

Consider a planar graph $G(V, E)$, an edge-capacity function $c : E \to \mathbb{N}$, an edge-destruction cost function $d : E \to \mathbb{N}$, and a budget B. In MIN NETWORK INHIBITION ON PLANAR GRAPHS, a feasible solution is an attack strategy to the network, i.e., a function $\alpha : E \to [0, 1]$ such that, $\sum_{e \in E} \alpha(e)d(e) \leqslant B$. The objective is to minimize the capability left in the damaged network, i.e., the total capacity of a cut in G where the capacity c' of an edge e is defined as $c'(e) = \alpha(e)c(e)$.

If a vertex disappears, the track of α in the surviving network always verifies the definition of α. Furthermore, since for the edges that disappear with the absence of a vertex we can assume that their destruction-cost becomes 0, the quantity $\sum_{e \in E'} \alpha(e)d(e)$, where E' is the edge-set of the surviving graph, is always less than, or equal to, B. Hence, MS is feasible for MIN NETWORK INHIBITION ON PLANAR GRAPHS.

Consider an *a priori* solution α together with a cut C_α, and a subgraph $G[V']$ induced by a subset V' of V. Then:

$$
\begin{aligned}
m\left(G, C_\alpha\left(V'\right)\right) &= \sum_{(v_i, v_j) \in C_\alpha} \alpha\left(v_i, v_j\right) c\left(v_i, v_j\right) 1_{\{v_i \in V', v_j \in V'\}} \\
E\left(G, C_\alpha\right) &= \sum_{V' \subseteq V} \Pr\left[V'\right] \sum_{(v_i, v_j) \in C_\alpha} \left(\alpha\left(v_i, v_j\right) c\left(v_i, v_j\right)\right. \\
&\quad \left. \times 1_{\{v_i \in V', v_j \in V'\}}\right)
\end{aligned}
\qquad [7.11]
$$

Since the facts $v_i \in V'$ and $v_j \in V'$ are independent, [7.11] becomes:

$$
\begin{aligned}
E\left(G, C_\alpha\right) &= \sum_{(v_i, v_j) \in C_\alpha} \alpha\left(v_i, v_j\right) c\left(v_i, v_j\right) \sum_{V' \subseteq V} \left(1_{\{v_i \in V'\}} \times 1_{\{v_j \in V'\}}\right. \\
&\quad \left. \times \Pr\left[V'\right]\right) \\
&= \sum_{(v_i, v_j) \in C_\alpha} \alpha\left(v_i, v_j\right) c\left(v_i, v_j\right) p_i p_j
\end{aligned}
\qquad [7.12]
$$

Based upon [7.12], one concludes that the optimal *a priori* solution for PROBABILISTIC MIN NETWORK INHIBITION ON PLANAR GRAPHS is the optimal solution of a weighted version of MIN NETWORK INHIBITION ON PLANAR GRAPHS where except for functions c and d, an edge-weight function $w : E \rightarrow \mathbb{Q}$, $(v_i, v_j) \mapsto p_i p_j$, is also provided, and the objective becomes to minimize the weighted capability left in the damaged network, i.e., the quantity $\sum_{e \in C_\alpha} \alpha(e)c(e)w(e)$.

Let us note that this explicit analysis is performed due to the fact that this problem is not very well-known. In fact, MIN NETWORK INHIBITION ON PLANAR GRAPHS fits perfectly the conditions of Theorem 6.3 in Chapter 6, if one considers that a feasible solution for it is not only the specification of α but also the specification of a cut.

7.4.7. MIN VERTEX k-CUT

Consider a graph $G(V, E)$, an integer k, a subset $S = \{s_i, t_i : i = 1, \ldots, k\}$ of "special vertices" of V and a vertex-weight function $w : V \setminus S \rightarrow \mathbb{N}$. MIN VERTEX k-CUT consists of determining a vertex k-cut, i.e., a subset $C \subseteq V \setminus S$ of vertices such that their deletion from G disconnects any s_i from t_i for $i = 1, \ldots, k$, minimizing the quantity $\sum_{v \in C} w(v)$.

Strategy MS is not feasible if we allow that a vertex in S may be absent. However, it becomes feasible, if absence of any $v \in S$ is not allowed, i.e., if the presence probability of any $v \in S$ is equal to 1. We deal with this assumption. In this case, PROBABILISTIC MIN VERTEX k-CUT fits the conditions of Theorem 6.1 and hence is **NP**-hard.

7.4.8. MIN MULTI-WAY CUT

Consider a graph $G(V, E)$, a subset S of "terminal vertices" of V and an edge-weight function $w : V \setminus S \rightarrow \mathbb{N}$. MIN MULTI-WAY CUT consists of determining a multi-way cut, i.e., a set $E' \subseteq E$ such that the removal of E' from E disconnects any terminal from all the others, minimizing the quantity $\sum_{e \in E'} w(e)$.

Under the same hypotheses as for MIN VERTEX k-CUT (section 7.4.7) about the presence-probabilities of the terminal vertices, MS is feasible. Then, PROBABILISTIC MIN MULTI-WAY CUT fits the conditions of Theorem 6.3 and hence it is **NP**-hard.

7.4.9. MIN MULTI-CUT

Consider a graph $G(V, E)$, a set $S \subseteq V \times V$ of source-sink pairs and an edge-weight function $w : E \rightarrow \mathbb{N}$. MIN MULTI-CUT consists of determining a minimum-weight multi-cut, i.e., a set $E' \subseteq E$, such that the removal of E' from E disconnects s_i from t_i, for any pair $(s_i, t_i) \in S$.

Under the same hypotheses as for MIN VERTEX k-CUT and MIN MULTI-WAY CUT (sections 7.4.7 and 7.4.8, respectively) about the presence-probabilities of the vertices in S, MS is feasible. Then, PROBABILISTIC MIN MULTI CUT fits the conditions of Theorem 6.3 and hence it is **NP**-hard.

7.4.10. MIN RATIO-CUT

Consider a graph $G(V, E)$, an edge-capacity function $c : E \rightarrow \mathbb{N}$, k commodities, i.e., k pairs $(s_i, t_i) \in V \times V$ and a demand d_i for any pair (s_i, t_i). The objective for MIN RATIO-CUT is to compute a cut minimizing its total capacity divided by the demand across the cut, i.e., minimizing the quantity:

$$\frac{\sum\limits_{\substack{v_i \in V_1 \\ v_j \in V_2 \\ (v_i, v_j) \in E}} c(v_i, v_j)}{\sum\limits_{i:|\{s_i, t_i\} \cap V_1| = 1} d_i}$$

If the presence probabilities of the "distinguished vertices" s_i, t_i, $i = 1, \ldots, |S|$, are different from 1, i.e., if these vertices are not always present, then MS is not always feasible for MIN RATIO-CUT. In fact, there may exist induced subgraphs of G for which at least one of the two sets V_1 and V_2 induced by the *a priori* cut is empty. In this case, there does not exist a surviving cut for the subgraph considered.

Suppose now that s_i and t_i, $i = 1, \ldots, |S|$, have presence-probability 1, i.e., they are always present in any of the subgraphs of G. Assume also that any of the sets V_1 and V_2 contains at least one of the s_i, t_i, for some $i = 1, \ldots, |S|$ (otherwise, as above MS may be unfeasible). In this case MS is feasible for MIN RATIO-CUT. Indeed, note that a cut remains always a cut, even if some of the vertices of the graph disappear. Furthermore, if for any $i \in \{1, \ldots, |S|\}$, both s_i and t_i are in the same set V_1 or V_2, then the total demand is zero and such a solution has no sense for the problem considered. Note finally that, since for any $i = 1, \ldots, |S|$, s_i and t_i are always present, the total demand across the cut, i.e., the quantity $\sum_{i:|\{s_i,t_i\} \cap V_1|=1} d_i$ is independent from the finally present subgraph of G, i.e., it is a constant with respect to V' in the expression for the functional. All this makes that MIN RATIO-CUT fits the conditions of Theorem 6.3. Considering an *a priori* cut E', its functional can then be written as:

$$E(G, E') = \frac{\sum\limits_{\substack{v_i \in V_1 \\ v_j \in V_2 \\ (v_i, v_j) \in E}} c(v_i, v_j)\, p_i p_j}{\sum\limits_{i:|\{s_i, t_i\} \cap V_1| = 1} d_i}$$

7.4.11. MIN b-BALANCED CUT

Consider a graph $G(V, E)$, a vertex-weight function $w : V \rightarrow \mathbb{N}$, an edge-cost function $c : E \rightarrow \mathbb{N}$ and $b \in \mathbb{Q}$, such that $b \in (0, 1/2]$. A feasible solution for MIN b-BALANCED CUT is a vertex-cut, i.e., a subset $C \subseteq V$ such that $\min\{w(C), w(V \setminus C)\} \geqslant bw(C)$, where, for any subset $V' \subseteq V$, $w(V') = \sum_{v \in V'} w(v)$. The objective is to minimize the total cost of the cut between C and $V \setminus C$, i.e., the quantity $\sum_{e \in \delta(C)} c(e)$, where $\delta(C) = \{e = (v_i, v_j) : e \in E, v_i \in C, v_j \in V \setminus C\}$.

There might exist subgraphs $G[V']$ of G, induced by some sets $V' \subseteq V$, for which given an *a priori* solution C, either $C \cap V'$, or $(V \setminus C) \cap V'$ is empty. In this case, the feasibility constrained is violated since $0 \not\geqslant bw(C)$. Hence, MS is not always feasible for MIN b-BALANCED CUT.

7.4.12. MIN b-VERTEX SEPARATOR

Consider a graph $G(V, E)$ of order n, and a number $b \in \mathbb{Q}$, such that $b \in (0, 1/2]$. A feasible solution for MIN b-VERTEX SEPARATOR is a partition of V into three sets A, B and C, such that $\max\{|A|, |B|\} \leqslant bn$, and no edge in E has one of its endpoints in A and another one in B; C is then called a *separator*. The objective is to determine a minimum-size such separator.

Absence of a vertex from the set C of an *a priori* solution for MIN b-VERTEX SEPARATOR reduces n, leaving $|A|$ and $|B|$ unchanged. This could violate the constraint on the quantity $\max\{|A|, |B|\}$. So, MS is not always feasible for MIN b-VERTEX SEPARATOR.

7.4.13. MIN QUOTIENT CUT

Consider a graph $G(V, E)$, a vertex-weight function $w : V \rightarrow \mathbb{N}$ and an edge-cost function $c : E \rightarrow \mathbb{N}$. In MIN QUOTIENT CUT, the objective is to determine a set $C \subseteq V$ minimizing the quantity $c(C)/\min\{w(C), w(V \setminus C)\}$, where $w(V') = \sum_{v \in V'} w(v)$, and $c(C) = \sum_{(u,v) \in E \text{ and } (u \in C, v \in V \setminus C) \text{ or } (u \in V \setminus C, v \in C)} c(u, v)$.

If, given an *a priori* solution C and a present subset V' of V, because of absence of some vertices either, $C \cap V' = \emptyset$, or $(V \setminus C) \cap V' = \emptyset$, then $\min\{w(C), w(V \setminus C)\} = 0$ and no value can be computed for the result of application of MS on C. Hence, this strategy is not feasible for MIN QUOTIENT CUT.

7.4.14. MIN k-VERTEX CONNECTED SUBGRAPH

Consider a connected graph $G(V, E)$ and a constant $k \geqslant 2$. In MIN k-VERTEX CONNECTED SUBGRAPH, the objective is to determine a k-vertex connected spanning partial subgraph $G'(V, E')$ of G, i.e., a spanning partial subgraph of G which cannot be disconnected by removing less than k vertices, with the least possible number of edges, i.e., with $|E'|$ as small as possible.

Observe that the absence of some vertices may disconnect G. So, MS is not always feasible for MIN k-VERTEX CONNECTED SUBGRAPH.

7.4.15. MIN k-EDGE CONNECTED SUBGRAPH

This is the same problem as MIN k-VERTEX CONNECTED SUBGRAPH (discussed in section 7.4.14), modulo the fact that the removal constraint carries over the edges. The result is the same as in section 7.4.14.

7.4.16. MIN BICONNECTIVITY AUGMENTATION

Consider a graph $G(V, E)$ and a symmetric vertex-weight function $w : V \rightarrow \mathbb{N} \times \mathbb{N}$. A feasible solution for MIN BICONNECTIVITY AUGMENTATION is an augmenting set E' for G, i.e., a set E' of unordered pairs of vertices from V such that $G(V, E \cup E')$ is biconnected. The objective consists of determining an augmenting set E' minimizing its total weight, i.e., the quantity $\sum_{(u,v)\in E'} w(u, v)$.

Absence of a vertex from an *a priori* augmenting set E' entails the absence of the edges of E' incident to it and, in this case, biconnectivity might be broken. So, MS is not feasible for MIN BICONNECTIVITY AUGMENTATION.

7.4.17. MIN STRONG CONNECTIVITY AUGMENTATION

This is the same problem as MIN BICONNECTIVITY AUGMENTATION (as seen in section 7.4.16) up to the fact that G is an oriented graph. The result is the same as in section 7.4.16.

7.4.18. MIN BOUNDED DIAMETER AUGMENTATION

Consider a graph $G(V, E)$ and a positive integer $D \leqslant |V|$. In MIN BOUNDED DIAMETER AUGMENTATION, we search for a minimum-cardinality set E' of unordered pairs from $V \times V$ such that $G(V, E \cup E')$ has diameter (see section A.2 of Appendix A) D.

Note that the absence of some vertices from a graph changes its diameter in a non-monotonous way. Hence MS is not always feasible for MIN BOUNDED DIAMETER AUGMENTATION.

Appendix A

Mathematical Preliminaries

A.1. Sets, relations and functions

This section recalls basic elements from sets, relations and functions. For further information, the interested reader can be referred to [LEW 81].

A set S is a collection of distinct elements of some universe $\mathcal{U}$. We use notation $S = \{a, b, c, d\}$ to represent a set S defined on elements a, b, c and d. This is an explicit representation of S. An implicit representation of a set S is to define it with respect to some property verified by all of its elements and only by them. This compact representation is very useful to denote sets with an infinite number of elements. A set with a finite number of elements is called *finite*; otherwise, it is called *infinite*. The *cardinality* of a set S (denoted by $|S|$) is the number of its elements.

The symbols $\in$ and $\notin$ denote the fact that an element belongs, or does not belong to a set, respectively. Dealing with set $S = \{a, b, c, d\}$, $a \in S$ and $z \notin S$.

Particularly interesting sets are:

– $\mathbb{N}$, the set of natural numbers and $\mathbb{N}^+$, the set of positive natural numbers;

– $\mathbb{Z}$, the set of integers and $\mathbb{Z}^+$, the set of positive integers;

– $\mathbb{Q}$, the set of rational numbers;

– $\mathbb{R}$, the set of real numbers and, more generally, $\mathbb{R}^m$, the set of real points in an m-dimensional space.

Two sets S and S' are equal (denoted by $S = S'$), if any element of S is also an element of S' and vice versa; otherwise, S and S' are not equal (denoted by $S \neq S'$). A set S' is a subset of S (denote by $S' \subseteq S$), if any element of S' is also an element of S; S' is a proper subset of S (denoted $S' \subset S$), if $S' \subseteq S$ and $S' \neq S$. Two sets are called *disjoint* if they do not include common elements.

A set S with $|S| = 1$ is called a *singleton*. In other words, a singleton $S = \{a\}$ includes a single element a. A set S is *empty* (denoted by $S = \emptyset$) if it contains no element; otherwise, S is called non-empty. The cardinality of an empty set is 0.

Given two sets X and Y:

– their *union*, denoted by $X \cup Y$, is the set of all elements that belong either to X, or to Y; the union of more than two sets $X_1, X_2, \ldots, X_k$, is denoted by $\cup_{i=1}^{k} X_i$; the union of an infinite number of sets $X_1, X_2, \ldots$, will be denoted by $\cup_{i=1}^{\infty} X_i$;

– their *intersection*, denoted by $X \cap Y$, is the set of all elements common to both X and Y;

– the difference X from Y, denoted by $X \setminus Y$ is the set of all elements of X that do not belong to Y; the difference $Y \setminus X$ is defined analogously;

The following basic equalities hold:

$$|X \cup Y| = \begin{cases} |X| + |Y| - |X \cap Y| & \text{if } X \text{ and } Y \text{ are not disjoint} \\ |X| + |Y| & \text{if } X \text{ and } Y \text{ are disjoint} \end{cases}$$

$$|X \setminus Y| = \begin{cases} |X| - |X \cap Y| & \text{if } X \text{ and } Y \text{ are not disjoint} \\ |X| & \text{if } X \text{ and } Y \text{ are disjoint} \end{cases}$$

If X, Y and Z are sets, then the following laws hold.

***Idempotency*:** $X \cup X = X$ and $X \cap X = X$.

***Commutativity*:** $X \cup Y = Y \cup X$ and $X \cap Y = Y \cap X$.

***Associativity*:** $(X \cup Y) \cup Z = X \cup (Y \cup Z)$ and $(X \cap Y) \cap Z = X \cap (Y \cap Z)$.

***Distributivity*:** $X \cup (Y \cap Z) = (X \cup Y) \cap (X \cup Z)$ and $X \cap (Y \cup Z) = (X \cap Y) \cup (X \cap Z)$.

***Absorption*:** $X \cap (X \cup Y) = X$ and $X \cup (X \cap Y) = X$.

***De Morgan's laws*:** $X \setminus (Y \cup Z) = (X \setminus Y) \cap (X \setminus Z)$ and $X \setminus (Y \cap Z) = (X \setminus Y) \cup (X \setminus Z)$.

The *power-set* of a set S (denoted by 2^S) is the set of all the subsets of S (including $\emptyset$ and S itself). If S is finite with cardinality n, then $|2^S| = 2^{|S|} = 2^n$.

Given a non-empty set S, a *partition* on S is a subset $\mathcal{S}$ of 2^S such that every element of $\mathcal{S}$ is non-empty, the elements of $\mathcal{S}$ are pairwise disjoint, and the union of its elements is equal to S.

Given a set S, the elements of which verify some property π, S is said to be *maximal for the inclusion with respect to* π if any addition of some other element in S results in a set that does not verify π. It is said to be *minimal for the inclusion with respect to* π if any removal of some element of S results is a set that does not verify π. Whenever property π is well understood by the context, we use terms *maximal* and *minimal* set, respectively.

A *sequence* of elements is a list of these elements in some order. A sequence is expressed by writing a list of its elements between parentheses. A sequence can be infinite or finite; finite sequences are called *tuples*. A tuple on k elements is called a *k-tuple*. Frequently, a 2-tuple is called a *pair* and a 3-tuple a *triple*.

The *Cartesian product* of k sets $S_1, \ldots, S_k$, denoted by $S_1 \times \ldots \times S_k$, is the set of all k-tuples $(s_1, \ldots, s_k)$ with $s_i \in S_i$, $i = 1, \ldots, k$. If, for any $i = 1, \ldots, k$, $S_i = S$, then the Cartesian product is denoted by S^k; $|S^k| = |S|^k$.

A k-dimensional vector $\vec{v}$ over a set S is an element of S^k. The i-th component of $\vec{v}$ will be denoted by v_i; in other words, $\vec{v} = (v_1, v_2, \ldots, v_k)$.

Given two sets X and Y, any subset $R \subseteq X \times Y$ is called a *binary relation* between X and Y. The *domain* of R is the set of all x, such that $(x, y) \in R$, for some $y \in Y$. Analogously, the *range* of R is the set of all y such that $(x, y) \in R$, for some $x \in X$. A binary relation R between X and X itself is called a *binary relation in* X. Consider such a relation R. Then:

– R is *reflexive*, if $(x, x) \in R$, for any $x \in X$;

– R is *symmetric*, if, for all pairs of elements x and y $(x, y) \in R \Rightarrow (y, x) \in R$;

– R is *transitive*, if, for all triples (x, y, z) $(x, y) \in R \wedge (y, z) \in R \Rightarrow (x, z) \in R$.

A binary relation in X that is simultaneously reflexive, symmetric and transitive is called an *equivalence relation*.

Given two sets X and Y, a *function* ϕ from X to Y, denoted by $\phi : X \rightarrow Y$, is a binary relation between X and Y that includes, at most, one pair (x, y) for any $x \in X$. A function with domain X and range Y is called a function *from X onto Y*. For a function $\phi : X \rightarrow Y$ and a pair (x, y) belonging to ϕ, we shall write $\phi(x) = y$ (or, sometimes, $\phi : x \mapsto y$) instead; x is called the *argument* and y the *value* of ϕ.

Consider a function $\phi : X \rightarrow Y$. Then:

– ϕ is *total* if its domain coincides with X;

– ϕ is *partial* if its domain is a subset of X;

– ϕ is *many-to-one* if the value of ϕ coincides on more than one distinct arguments;

– ϕ is *one-to-one*, or *injective*, if, for all $x, x' \in X$ with $x \neq x'$, $\phi(x) \neq \phi(x')$;

– ϕ is *surjective*, if Y coincides with the range of ϕ;

– ϕ is *bijective* if it is both injective and surjective; a bijective function is very frequently called a *bijection*.

Given two functions $\phi, \psi : \mathbb{N} \rightarrow \mathbb{N}$:

– $\phi(n) = O(\psi(n))$, if there exist constants κ, κ' and n_0 such that, for all $n \geqslant n_0$, $\phi(n) \leqslant \kappa\psi(n) + \kappa'$;

– $\phi(n) = \Omega(\psi(n))$, if there exist constants κ, κ' and n_0 such that, for all $n \geqslant n_0$, $\phi(n) \geqslant \kappa\psi(n) + \kappa'$;

– $\phi(n) = \Theta(\psi(n))$ if both $\phi(n) = O(\psi(n))$ and $\phi(n) = \Omega(\psi(n))$ hold;

– $\phi(n) = o(\psi(n))$, if $\lim_{n\rightarrow\infty}(\phi(n)/\psi(n) = 0$.

Binary relations and functions can be easily generalized into more than two ground sets. Given $k+1$ sets $X_1, \ldots, X_k, Y$, a *k-ary relation* between $X_1, \ldots, X_k$ is any subset $R \subsetneq X_1 \times \ldots \times X_k$. A *k-ary function* $f : X_1 \times \ldots \times X_k \rightarrow Y$ is a $(k+1)$-ary relation between the sets $X_1, \ldots, X_k, Y$ that includes, at most, one $(k+1)$-tuple $(x_1, \ldots, x_k, y)$ for any $(x_1, \ldots, x_k) \in X_1 \times \ldots \times X_k$.

Given a function $f : X \rightarrow Y$, f admits an *inverse function*, denoted by $f^{-1} : Y \rightarrow X$, if the following holds: $f(x) = y \Leftrightarrow f^{-1}(y) = x$.

Consider two sets X and Y; they are called *equinumerous* if there exists a bijection $\phi : X \rightarrow Y$. In general, if a set is finite then, if its cardinality is n, it is equinumerous[1] with $\{1, 2, \ldots, n\}$. An infinite set S is called **countably infinite** if it is equinumerous with $\mathbb{N}$; it is called *countable* if it is finite, or countable infinite; finally, it is called *uncountable* if it is not countable.

We denote by $1_{\{X\}}$ the indicator function of the fact X, i.e.:

$$1_{\{X\}} = \begin{cases} 1 & \text{if } X \text{ is true} \\ 0 & \text{otherwise} \end{cases}$$

A.2. Basic concepts from graph-theory

We give in this section some basic elements of graph theory. For more details and fundamental concepts, the interested reader can be referred to [BER 73, BOL 79].

1. The empty set is finite and equinumerous with itself.

An *undirected graph* $G(V, E)$ is a pair of finite sets (V, E), such that E is a binary symmetric relation in V. The set V is the set of *vertices* (sometimes called *nodes*) and the set E is the set of edges. By the symmetry of E, an edge (u, v) identifies both the pair (u, v) and the pair (v, u). For a $(u, v) \in E$, u and v are said to be *adjacent*, or *neighbors*; they are also called the *endpoints* of edge (u, v). This edge is said to be *incident* to v and u. For a vertex $v \in V$, we denote by $\Gamma(v)$ the set of its neighbors.

The *degree* of a vertex $v \in V$ is the number of vertices adjacent to it (the number $|\Gamma(v)|$ of its neighbors). The degree of G, denoted by $\Delta(G)$, or simply by Δ when no confusions arises, is the degree of its maximum-degree vertex; in other words, $\Delta(G) = \max_{v_i \in V}\{|\Gamma(v_i)|\}$. The degree of a minimum-degree vertex of G will be denoted by $\delta(G)$ (or simply by δ when no confusion arises); in other words, $\delta(G) = \min_{v_i \in V}\{|\Gamma(v_i)|\}$. A graph is said to be *d-regular*, if all its vertices have degree d, i.e., $\Delta(G) = \delta(G) = d$.

A *directed graph* $G(V, A)$ is a pair of finite sets (V, A), such that A is a binary relation in V, not necessarily symmetric. The elements of A are called *arcs*. For an arc (u, v), u is said *initial endpoint* of (u, v) and v is said *final endpoint* of (u, v).

A graph $G'(V', E')$ is said to be *subgraph of* $G(V, E)$ *induced by* V', if $V' \subseteq V$ and $E' = \{(u, v) : u, v \in V' \wedge (u, v) \in E\}$. A graph $G'(V', E'')$ is said to be *partial subgraph of* $G(V, E)$, if $V' \subseteq V$ and $E'' \subseteq E'$.

A *weighted graph* $G(V, E, w)$ is a graph $G(V, E)$ together with a function $w : V \rightarrow \mathbb{Q}$, or $w : E \rightarrow \mathbb{Q}$, that associates a weight with any vertex, or any edge (or any arc) of G. Sometimes function w is represented as a vector $\vec{w}$, the components of which are the values of function w on the elements of V or of E. For any weighted graph $G(V, E, w)$, and for any vertex $v \in V$ (resp., any edge $(u, v) \in E$), $w(v)$ (resp., $w(u, v)$) is called the weight of v (resp., of (u, v)).

A graph $G(V, E)$ is called *complete*, or *clique* if $E = V^2$, or if $E = V^2 \setminus \{(v, v) : v \in V\}$). In other words, any two vertices of a clique are adjacent. In the former case G will be called complete (or clique) with loops. A complete graph on n vertices verifies $|E| = n(n-1)/2$.

Given a graph $G(V, E)$ and two vertices v_1 and v_k, an *elementary path* from v_1 to v_k is a sequence of distinct vertices $(v_1, v_2, \ldots, v_k)$ such that, for $i = 1, \ldots, k-1$, $(v_i, v_{i+1}) \in E$. The definition of a path for a directed graph $G(V, A)$ is similar up to the fact that, for $i = 1, \ldots, k-1$, the pair (v_i, v_{i+1}) belongs to A. In this latter case the path $(v_1, v_2, \ldots, v_k)$ is called a *directed path* (or simply a path, when no confusion arises). Sometimes, a path is represented as sequence of its edges (or arcs). The *length* of a path is the number of its edges. When dealing with edge-weighted graphs, the value of the path is the sum of the weights of its edges (or arcs). A *cycle* of G is a path

$(v_1, v_2, \ldots, v_k)$ with $v_1 = v_k$. If a cycle contains all the vertices of G, then it is called a *Hamiltonian cycle*.

The *diameter* of a graph is the maximum distance (i.e., the length of the longest elementary path) of any pair of its vertices.

A graph is connected if, for any pair of distinct vertices u and v, there exists a path from u to v; otherwise the graph is said to be *non-connected.* A connected component of a non-connected graph G is some connected subgraph of G. A graph is said to be biconnected if, for any pair of distinct vertices u and v, there exist two vertex-disjoint paths from u to v. A directed graph is said to be strongly connected if, for any pair of distinct vertices u and v, there exist two vertex-disjoint directed paths from u to v.

A *tree* $T(V, E)$ is a connected graph with no cycles. It is *rooted* if there exists one node designated as the *root*. In this book we deal only with rooted trees. A common way to recursively define a rooted tree is the following:

– a single vertex v is a tree rooted at v;

– assume that v is a vertex and $T_1, \ldots, T_k$ are trees with roots $v_1, \ldots, v_k$, respectively; a new tree can be obtained by linking v with $v_1, \ldots, v_k$; in this new tree v is the root and $v_1, \ldots, v_k$ are its *children*.

The *height* of a tree-node v is the length of the path from the root of the tree to v. The *height* of a tree is the maximum over the heights of its nodes. In a tree, a vertex with no children is called a *leaf.* In any tree $T(V, E)$, $|E| = |V| - 1$. A usual (although somewhat informal) way to characterize a tree is the following: "a tree is a connected acyclic graph such that if one adds an edge, then one creates a cycle and if one removes an edge, then one disconnects the graph". Following this characterization, *a tree is a graph that, in terms of edges, is minimal for the connectivity and maximal for the acyclicity*. A rooted tree of height 1 will be called a *star*.

A directed graph $G(V, A)$ is called *acyclic*, if it contains no *directed cycle*.

A directed graph $G(V, A)$ is called *transitive*, if for any pair $(u, v) \in V \times V$ the following condition holds: if there exists $w \in V$ such that $(u, w) \in A$ and $(w, v) \in A$, then $(u, v) \in A$.

An undirected graph is called a *comparability graph* if its edges can be oriented in such a way that the resulting directed graph is acyclic and transitive.

The *complement*[2] of a graph $G(V, E)$ is the graph $\bar{G}(V, E')$ with $E' = \{(V \times V) \setminus (E \cup \{(v, v) : v \in V\})\}$.

2. Sometimes called a *complementary graph*.

A graph $G(V, E)$ is *bipartite* if its vertex-set can be partitioned into two subsets U and D such that any edge in E links a vertex of U to a vertex of D.

A graph is called *chordal* if any cycle of length at least 4 has a chord (i.e., an edge linking two vertices of the cycle).

A graph is called *planar* if it is possible to represent it on a plane in which the vertices are distinct points, the edges simple curves and no two edges cross one another.

A graph $G(V, E)$ is called a *split graph* if its vertex-set V can be partitioned into two subsets V_1 and V_2 such that, one of them, say V_1, is an independent set and the other one, V_2 is such that $G[V_2]$ is a clique.

We now define some characteristic graph-configurations seen in this book.

Given a graph $G(V, E)$, an *independent set* is a subset $V' \subseteq V$ such that, for any pair $(u, v) \in V' \times V'$, $(u, v) \notin E$. In other words, an independent set is a set of pairwise non-adjacent vertices. When dealing with (vertex-)weighted graphs, the measure of an independent set V' is the sum of the weights of the vertices of V', unless it is defined otherwise. The cardinality of a maximum independent set of G, also called *independence number* or *stability number*, is denoted by $\alpha(G)$.

Given a graph $G(V, E)$, a *clique* is a subset $V' \subseteq V$ such that, for any pair $(u, v) \in V' \times V'$, $(u, v) \in E$. In other words, a clique is a set of pairwise adjacent vertices.

Given a graph $G(V, E)$, a *vertex cover* is a subset $V' \subseteq V$ such that, for any $(u, v) \in E$, either u or v belongs to V'. When dealing with (vertex-)weighted graphs, the measure of a vertex cover V' is the sum of the weights of the vertices of V'.

The following relations link, for a fixed graph $G(V, E)$, the three configurations defined just above:

1) for any independent set V' of G, $V \setminus V'$ is a vertex cover for G;

2) any independent set V' of G is a clique in $\bar{G}$.

Given a graph $G(V, E)$, a *coloring* is a function from some color-set onto V such that no adjacent vertices in G receive (that is, are colored with) the same color. In other words the vertices of G colored with the same color form an independent set. Then, a coloring of G (also called coloring of V) is a partition of V into independent sets. The cardinality of a minimum coloring of G is called *a chromatic number* of G and is denoted by $\chi(G)$.

Given a graph $G(V, E)$, a *matching* M is a subset of E such that no two edges in M share the same endpoint. In other words if we represent an edge as a set of its endpoints, a matching is a set of pairwise disjoint edges. A matching M is said to be *perfect* if it saturates all the vertices of G, i.e., if any $v \in V$ is the endpoint of some edge of M.

Consider an oriented graph $G(V, A)$ and assume that G contains at least a vertex, denoted by s, such that s is the initial endpoint of any arc incident to it, and at least a vertex, denoted by t, such that t is the final endpoint of any arc incident to it. Vertices s and t are called *source* and *sink*, respectively. Fix a pair (s, t). Assume also that with any arc $a_i = (u, v) \in A$ are associated two numbers b_i and c_i in $\mathbb{Q}$ such that, for any $i = 1, \ldots, |A|$, $-\infty \leqslant b_i \leqslant c_i \leqslant +\infty$. Denote by $\vec{b}$ the vector of b_is and by c the vector of c_is, $i = 1, \ldots, |A|$. Such a graph is usually denoted by $G(V, A, s, t, \vec{b}, \vec{c})$ and is called a *network*. Whenever $b_i = 0$, for any $i = 1, \ldots, |A|$, the network G is called *transportation network*. A flow of a network G is a vector $\vec{\phi} = (\phi_1, \ldots, \phi_{|A|})$, $\phi_i \in \mathbb{Q}$ such that:

1) for any $i \in \{1, \ldots, |A|\}$, $b_i \leqslant \phi_i \leqslant c_i$;

2) for any vertex $v \in V \setminus \{s, t\}$, the sum of the arc-flows entering v equals the sum of the arc-flows leaving v, i.e., $\sum_{a_i \in E(v)} \phi_i = \sum_{a_i \in L(v)} \phi_i$, where $E(v) = \{(x, v) \in A\}$ and $L(v) = \{(v, x) \in A\}$. This property is a kind of conservation law called *Kirchoff's law*.

The maximum flow problem can simply be specified as follows: consider a "fictitious" arc a_0 from t to s, set $b_0 = -\infty$ and $c_0 = +\infty$, denote by G' the so-modified network and determine a flow $\vec{\phi} = (\phi_0, \phi_1, \ldots, \phi_{|A|})$ such that:

1) for any $i \in \{0, \ldots, |A|\}$, $b_i \leqslant \phi_i \leqslant c_i$;

2) for any vertex $v \in V$, the sum of the arc-flows entering v equals the sum of the arc-flows leaving v, i.e., $\sum_{a_i \in E(v)} \phi_i = \sum_{a_i \in L(v)} \phi_i$;

3) ϕ_0 is maximized.

Given a graph G, an *edge covering* E' is a subset of E such that any vertex $v \in V$ is the endpoint of an edge in E'.

Given a graph G and a partition of V into two sets V_1 and V_2, a *cut* is a set of edges in E having one of their endpoints in V_1 and the other one in V_2.

A.3. Elements from discrete probabilities

An *event space*, or *probability space*, Ω is a set of elements (also called *elementary events*) ω, such that $\cup_\omega \{\omega\} = \Omega$. With any $\omega \in \Omega$, we associate a probability $\Pr[\omega] \in [0, 1]$, such that $\sum_{\omega \in \Omega} \Pr[\omega] = 1$. The set $\{\Pr[\omega] : \omega \in \Omega\}$ is called *probability law* or *probability distribution*.

An *event* A is a subset of Ω. The probability of A is defined as: $\Pr[A] = \sum_{\omega \in A} \Pr[\omega]$.

A *random variable* (r.v. for short) is a function defined on the elementary events of some probability space. Any random variable X can be characterized by the probability distribution $\{\Pr[X = \omega]\}$ of its values.

Two random variables X and Y are called *independent* if and only if: $\Pr[X = x \wedge Y = y] = \Pr[X = x] \cdot \Pr[Y = y]$.

The *expectation* $E(X)$ of a random variable X (representing, in some sense, the mean value of X) is defined as:

$$E(X) = \sum_{x \in X(\Omega)} x \cdot \Pr[X = x]$$

where $X(\Omega)$ is the set of all the values the r.v. X can be assigned.

PROPERTY A.1.– Let X and Y be two random variables over a probability space Ω and let $a \in \mathbb{R}$. Then:

– $E(X + Y) = E(X) + E(Y)$;

– $E(aX) = aE(X)$;

– if X and Y are independent, then $E(X \cdot Y) = E(X) \cdot E(Y)$.

Given a random variable X, its *variance* $\mathrm{Var}(X)$ (representing the distribution of X around $E(X)$) is defined by:

$$\mathrm{Var}(X) = E\left((X - E(X))^2\right)$$

The *standard deviation* of an r.v. X is defined by: $\sigma(X) = \sqrt{\mathrm{Var}(X)}$. It represents the mean deviation between X and $E(X)$. If $\sigma(X)$ is small, then the value of X is almost always close to $E(X)$.

PROPERTY A.2.– Let X be an r.v. on a probability space Ω. Then:

– $\mathrm{Var}(X) = E\left(X^2\right) - (E(X))^2$;

– if Y is an r.v. and X and Y are independent, then $\mathrm{Var}(X + Y) = \mathrm{Var}(X) + \mathrm{Var}(Y)$.

Given two events A and B, the *conditional probability* of A subject to B, denoted by $\Pr[A|B]$, is defined by:

$$\Pr[A|B] = \frac{\Pr[A \wedge B]}{\Pr[B]}$$

If A and B are independent, then $\Pr[A|B] = \Pr[A]$.

Let $\Omega = \{\text{success}, \text{faillure}\}$ be a probabilistic space, with $\Pr[\text{success}] = p$ and $\Pr[\text{faillure}] = q = 1 - p$. Assume that an experiment is realized n times in Ω. Then, the random variable X that represents the total number of times (among the n realizations) where the experiment's result is "success", follows a probability law, called *binomial*, that is defined by:

$$\begin{aligned} \Pr(X = k) &= \binom{n}{k} p^k q^{n-k} \\ E(X) &= np \end{aligned}$$

The fact that X follows the binomial law is frequently denoted by $X \sim B(n, p)$.

Appendix B

Elements of the Complexity and the Approximation Theory

This chapter presents preliminary concepts from the complexity theory and the theory of the polynomial time approximation of **NP**-hard problems. For more details, see [AUS 99, COO 71, GAR 79, KAR 72, PAP 94, PAP 81, PAS 04].

B.1. Problem, algorithm, complexity

Informally, a problem can be seen as a specification of triple $(\mathcal{I}, \mathcal{C}, R)$ where:

– $\mathcal{I}$ is a set of instances;

– $\mathcal{C}$ is a set of properties defining the solutions of the problem;

– Q is some requirement about solutions; it can be:

- either a question about the existence of a solution satisfying $\mathcal{C}$; in this case we have what it is commonly called a *decision problem*;

- or a requirement for the computation of the best among the solutions satisfying $\mathcal{C}$; in this case we have what it is called *optimization (or search) problem.*

Let us note that for any optimization problem, there exists a decision counterpart. For instance, if we deal with a maximization problem, its decision version can be stated as follows: "given $(\mathcal{I}, \mathcal{C}, R)$ and a constant k, does there exist a solution verifying $\mathcal{C}$ with value at least k?".

An *algorithm* is a kind of virtual machine which, by performing elementary operations, correctly solves a problem. This, for a decision problem, means that the algorithm correctly answers *yes* if a solution exists that satisfies $\mathcal{C}$ and it answers *no*, otherwise; for an optimization problem, this means that the algorithm really computes the best among the solutions satisfying $\mathcal{C}$. Note also that for an optimization problem, any solution satisfying $\mathcal{C}$ (even not optimal) is called *feasible*.

The most common (although not the only) measure which we use here for the premium quality of an algorithm is its *time-complexity*, called simply *complexity* throughout this book, expressed as the number of the elementary operations performed by this algorithm. It is expressed as function of the size of the instance where the algorithm runs; the smaller the complexity, the better the algorithm supposed to solve the problem.

For any problem, its *complexity* is the complexity of the best (fastest) known algorithm solving it.

B.2. Some notorious complexity classes

From a computational point of view, the most natural problems can be classified into two main categories:

1) the *polynomial problems*; these are problems admitting algorithms solving them on an instance of size[1] n, with complexity $O(n^k)$ for some k independent of n; they form the complexity class **P**;

2) the *non-polynomial problems*; these are problems for which the best algorithms known have super-polynomial complexity, i.e., complexity in $O(f(n)^{g(n)})$, where f and g are functions increasing with n and $\lim_{n\to\infty} g(n) = \infty$.

A very well known subclass fitting Item 2 above is the class of problems solved with complexity $O(2^{n^k})$, for some k independent of n. All these problems form the class **EXPTIME** of exponential problems.

Dealing with decision problems[2], the most notorious complexity class (except **P**) is the class **NP**. Informally, this is the class of problems any solution of which is decidable in polynomial time. This means that given an instance of an **NP** problem, if an oracle guesses a solution, then there exists an algorithm that can decide, in polynomial time, if this solution satisfies $\mathcal{C}$. Obviously, $\mathbf{P} \subseteq \mathbf{NP}$ but no-one knows yet if this inclusion is strict or not. It is strongly conjectured that $\mathbf{P} \subset \mathbf{NP}$. This conjecture is one of the most fundamental conjectures of the complexity theory and the problem of proving or invalidating it is one of the most famous open scientific problems.

1. The size of an instance is the number of objects, under a specific encoding, needed for describing the instance in the computer.
2. The complexity theory is fundamentally based upon decision problems.

The notion of the class **NP** can be extended to also fit optimization problems. An optimization problem belongs to the class **NPO** if and only if its decision counterpart belongs to **NP**. In other words, **NPO** problems are the ones whose decision counterparts are in **NP**. Informally, an optimization problem belongs to **NPO**, if the following three conditions are satisfied:

1) feasibility of any solution is decidable in polynomial time;

2) the value of any feasible solution is computable in polynomial time;

3) at least one feasible solution is computable in polynomial time.

A more formal definition for this class is given in Chapter 1 (Definition 1.1).

B.3. Reductions and NP-completeness

The notion of the *polynomial reduction* is the most powerful tool for establishing relations among the complexities of different problems. Basically, a *reduction* from a problem Π to a problem Π' describes a strategy for solving Π' using an algorithm for Π. Informally, in this case Π' can be characterized as at least as algorithmically difficult as Π (since with an algorithm for the latter, one solves not only this one but also the former).

Two reductions deal mainly with problems in **NP** (and their optimal solution): the *Karp-reduction* (or *polynomial transformation*) and the *Turing-reduction*. Their formal definitions can be found in [AUS 99, GAR 79, PAP 94, PAP 81, PAS 04]. What is actually used in the litterature for proving **NP**-completeness is an intermediate form between these two reductions which can broadly sketched as follows. Consider two problems Π_1 and Π_2 and denote by $\mathcal{I}_{\Pi_1}$ and $\mathcal{I}_{\Pi_2}$ the sets of their instances, and by Sol_{Π_1} and Sol_{Π_2} the sets of their (feasible when dealing with optimization problems) solutions, respectively. A reduction is a pair (f, g) of functions, computable in polynomial time, such that:

– $f : \mathcal{I}_{\Pi_1} \rightarrow \mathcal{I}_{\Pi_2}$; for any $I \in \mathcal{I}_{\Pi_1}$, $f(I) \in \mathcal{I}_{\Pi_2}$;

– $g : \text{Sol}_{\Pi_2} \rightarrow \text{Sol}_{\Pi_1}$; for any $(I, S) \in \mathcal{I}_{\Pi_1} \times \text{Sol}_{\Pi_2}(f(I))$, $g(I, S) \in \text{Sol}_{\Pi_1}(I)$.

Then on the hypothesis that there exists an algorithm $\mathtt{A}$ for Π_2, algorithm $f \circ \mathtt{A} \circ g$ is an algorithm for Π_1.

The notion of the polynomial reduction turns out to be particularly relevant when dealing with **NP**, since it makes it possible to draw a subclass of it including the "hardest" among its problems. These are the famous **NP**-complete problems. A problem Π is **NP**-complete with respect to some polynomial reduction R if any problem $\Pi' \in$ **NP** R-reduces to Π. Broadly speaking, in order to prove that a particular problem Π_0 is **NP**-complete, it suffices (using the transitivity of the reduction sketched above) to prove the following two statements:

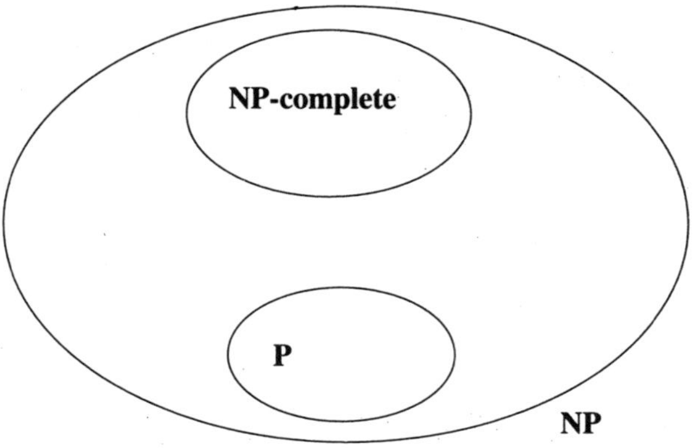

Figure B.1. *P, NP and NP-complete, under the assumption P ≠ NP*

1) $\Pi_0 \in$ **NP**;

2) there exists an **NP**-complete problem Π' such that Π' R-reduces to Π.

Transitivity of the reductions considered concludes that, if statements 1 and 2 hold, then any problem in **NP** reduces to Π_0 and, consequently, Π_0 is **NP**-complete.

Figure B.1 illustrates the landscape of problems in **NP**, under the commonly admitted conjecture: **P** $\neq$ **NP**. The class **NP** $\setminus$ (**P** $\cup$ **NP**-complete) is non-empty under the Karp-reduction ([LAD 75]).

Note finally that the notion of **NP**-completeness, dealing with decision problems, can be extended to fit optimization problems. An optimization problem is said to be **NP**-hard if its decision counterpart is **NP**-complete.

An intuitive way of thinking of **NP**-complete problems is the following one: no polynomial algorithm is known until now for these problems, but if such an algorithm were devised for just one of them, then it would be used as a subroutine for devising polynomial algorithms for any **NP**-complete problem. It is very widely believed that such a polynomial algorithm will never be designed.

B.4. Approximation of NP-hard problems

Since the general belief is that **NP**-hard problems will never be solved in polynomial time, if one wishes to devise fast algorithms for any of them, then one has to develop polynomial time methods providing, for any instance, feasible solutions, the values of which are as close to the values of the optimal solutions. These algorithms are called *polynomial time approximation algorithms*.

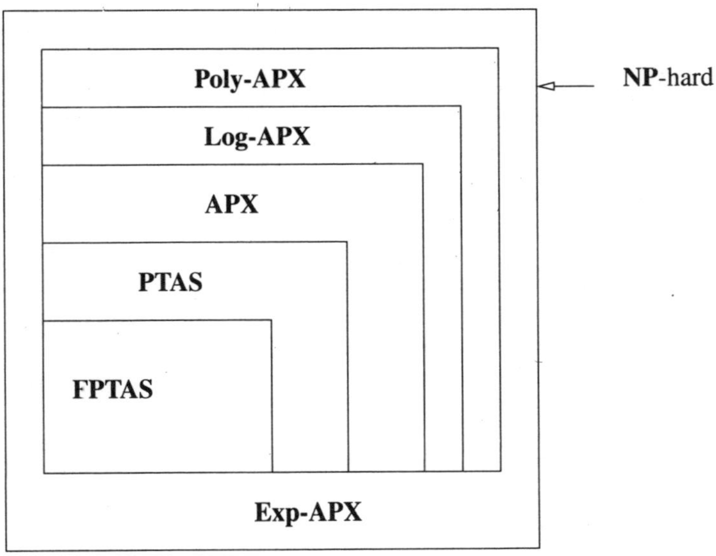

Figure B.2. *The main approximability classes, under the assumption* $\mathbf{P} \neq \mathbf{NP}$

Given an instance I of an **NPO** problem $\Pi = (\mathcal{I}, \mathrm{Sol}, m, \mathrm{goal})$, denote by $\mathrm{opt}(I)$ the optimal value (i.e., the value of an optimal solution) of I and by $m_{\mathtt{A}}(I, S)$ the value of a feasible solution S computed on I by a polynomial time approximation algorithm $\mathtt{A}$ solving Π. The "goodness" of $\mathtt{A}$ on I is expressed by the approximation ratio $\rho_{\mathtt{A}}(I, S) = m_{\mathtt{A}}(I, S)/\,\mathrm{opt}(I)$. The approximation ratio of $\mathtt{A}$ for Π is defined as:

$$\rho_{\mathtt{A}} = \begin{cases} \sup\limits_{I \in \mathcal{I}} \{r : \rho_{\mathtt{A}}(I, S) > r\} & \mathrm{goal} = \max \\ \inf\limits_{I \in \mathcal{I}} \{r : \rho_{\mathtt{A}}(I, S) < r\} & \mathrm{goal} = \min \end{cases}$$

Finally, the approximation ratio for Π, denoted by ρ_{Π}, is the approximation ratio of the best approximation algorithm known for Π; formally:

$$\rho_{\Pi} = \begin{cases} \max\{\rho_{\mathtt{A}} : \mathtt{A} \text{ an approximation algorithm for } \Pi\} & \mathrm{goal} = \max \\ \min\{\rho_{\mathtt{A}} : \mathtt{A} \text{ an approximation algorithm for } \Pi\} & \mathrm{goal} = \min \end{cases}$$

According to their approximation ratios, combinatorial optimization problems can be classified to approximability classes. The most notorious among them are the following ones:

Exp-APX: the class of problems approximable within ratios that are exponential (or the inverse of an exponential, when $\mathrm{goal} = \max$) with n, where n is the size of their instance;

Poly-APX: the class of problems approximable within ratio $O(n^{\epsilon})$ (or $O(n^{-\epsilon})$, when goal = max), for some $\epsilon > 0$;

Log-APX: the class of problems approximable within ratio $O(\log^{\epsilon} n)$ $(O(1/\log^{\epsilon} n)$, when goal = max);

APX: the class of problems approximable with constant ratios (independent of n and of any other instance-parameter);

PTAS: the class of problems approximable by *polynomial time approximation schemata*, that is, by sequences $\mathtt{A}_{\epsilon}$ of polynomial time approximation algorithms achieving, for any $\epsilon > 0$, approximation ratio $\rho_{\mathtt{A}_{\epsilon}} = 1 + \epsilon$ ($\rho_{\mathtt{A}_{\epsilon}} = 1 - \epsilon$ when goal = max) with complexity $O(n^k)$, where k is a constant not depending on n but eventually depending on $1/\epsilon$;

FPTAS: the class of problems approximable by *fully polynomial time approximation schemata*, i.e. by polynomial time approximation schemata that are polynomial with both n and $1/\epsilon$.

Obviously, polynomial time approximation schemata, *a fortiori* the full polynomial time approximation schemata, represent an "ideal" approximation behavior for an **NP**-hard problem.

Figure B.2 illustrates the approximation classes just defined. Dealing with them, the following relation holds:

$$\textbf{FPTAS} \subset \textbf{PTAS} \subset \textbf{APX} \subset \textbf{Log-APX} \subset \textbf{Poly-APX}$$

These inclusions are strict, unless **P** = **NP**.

Bibliography

[AHO 75] AHO A. V., HOPCROFT J. E., ULLMAN J. D., *The design and analysis of computer algorithms*, Addison-Wesley, Reading, MA, 1975.

[AUS 99] AUSIELLO G., CRESCENZI P., GAMBOSI G., KANN V., MARCHETTI-SPACCAMELA A., PROTASI M., *Complexity and approximation. Combinatorial optimization problems and their approximability properties*, Springer, Berlin, 1999.

[AVE 94] AVERBAKH I., BERMAN O., SIMCHI-LEVI D., "Probabilistic *a priori* routing-location problems", *Naval Res. Logistics*, vol. 41, p. 973–989, 1994.

[AVE 95] AVERBAKH I., BERMAN O., "Probabilistic sales-delivery man and sales-delivery facility location problems on tree", *Transportation Science*, vol. 29, num. 2, p. 184–197, 1995.

[BAA 78] BAASE S., *Computer algorithms*, Addison-Wesley, 1978.

[BAR 85] BAR-YEHUDA R., EVEN S., "A local-ratio theorem for approximating the weighted vertex cover problem", *Ann. Discrete Appl. Math.*, vol. 25, p. 27–46, 1985.

[BAR 88] BAR-NOY A., BELLARE M., HALLDÓRSSON M. M., SHACHNAI H., TAMIR T., "On chromatic sums and distributed resource allocation", *Information and Computation*, vol. 140, p. 183–202, 1988.

[BEL 57] BELLMAN R., *Dynamic programming*, Princeton University Press, Princeton, N.J., 1957.

[BEL 93] BELLALOUNA M., Problèmes d'optimisation combinatoires probabilistes, PhD Thesis, Centre de Recherche en Mathématiques Appliquées, Ecole Nationale des Ponts et Chaussées, Paris, France, 1993.

[BER 73] BERGE C., *Graphs and hypergraphs*, North Holland, Amsterdam, 1973.

[BER 79] BERENGUER X., "A characterization of linear admissible transformation for the m-traveling salesmen problem", *European J. Oper. Res.*, vol. 3, p. 232–249, 1979.

[BER 88] BERTSIMAS D. J., Probabilistic combinatorial optimization problems, PhD Thesis, Operations Research Center, MIT, Cambridge Mass., USA, 1988.

[BER 89] BERTSIMAS D. J., "On probabilistic traveling salesman facility location problems", *Transportation Sci.*, vol. 3, p. 184–191, 1989.

[BER 90a] BERTSIMAS D. J., "The probabilistic minimum spanning tree problem", *Networks*, vol. 20, p. 245–275, 1990.

[BER 90b] BERTSIMAS D. J., JAILLET P., ODONI A., "*A priori* optimization", *Oper. Res.*, vol. 38, num. 6, p. 1019–1033, 1990.

[BER 92] BERTSIMAS D. J., "A vehicle routing problem with stochastic demand", *Oper. Res.*, vol. 40, num. 3, p. 574–585, 1992.

[BER 93] BERTSIMAS D. J., HOWELL L. H., "Further results on the probabilistic traveling salesman problem", *European J. Oper. Res.*, vol. 65, p. 68–95, 1993.

[BER 96] BERTSIMAS D. J., SIMCHI-LEVI D., "A new generation of vehicle routing research: robust algorithms, addressing uncertainty", *Oper. Res.*, vol. 44, num. 2, p. 286–304, 1996.

[BJO 97] BJORNSSON Y., STEWART L., XU B., Comparability graph page, Available on `http://www.cs.ualberta.ca/~bing/GRAPH/comparability`, 1997.

[BOD 94] BODLAENDER H. L., JANSEN K., WOEGINGER G. J., "Scheduling with incompatible jobs", *Discrete Appl. Math.*, vol. 55, p. 219–232, 1994.

[BOL 79] BOLLOBÁS B., *Graph theory. An introductory course*, Springer-Verlag, New York, 1979.

[BOL 85] BOLLOBÁS B., *Random graphs*, Academic Press, London, 1985.

[BOU 84] BOURJOLLY J.-M., HAMMER P. L., SIMEONE B., "Node-weighted graphs having the König-Egervary property", *Math. Programming Stud.*, vol. 22, p. 44–63, 1984.

[CAI 92] CAI Y., KONG M. C., "Generation of all maximal cliques and related problems for certain perfect graphs", *Congr. Numer.*, vol. 90, p. 33–55, 1992.

[CHR 76] CHRISTOFIDES N., Worst-case analysis of a new heuristic for the traveling salesman problem, Technical Report num. 388, Grad. School of Industrial Administration, CMU, 1976.

[COO 71] COOK S. A., "The complexity of theorem-proving procedures", *Proc. STOC'71*, p. 151–158, 1971.

[DEM 99] DEMANGE M., PASCHOS V., Maximum-weight independent set is as "well-approximated" as the unweighted one, Cahier du LAMSADE num. 163, LAMSADE, Université Paris-Dauphine, June 1999.

[DEM 01] DEMANGE M., DE WERRA D., MONNOT J., PASCHOS V., Time slot scheduling of compatible jobs, Cahier du LAMSADE num. 182, LAMSADE, Université Paris-Dauphine, 2001, Available on `http://www.lamsade.dauphine.fr/cahdoc.html#cahiers`.

[DEM 02] DEMANGE M., DE WERRA D., MONNOT J., PASCHOS V., "Weighted node coloring: when stable sets are expensive", KUČERA L., Ed., *Proc. 28th International Workshop on Graph Theoretical Concepts in Computer Science, WG'02*, vol. 2573 of *Lecture Notes in Computer Science*, Springer-Verlag, p. 114–125, 2002.

[DIJ 59] DIJKSTRA E., "A note on two problems in connexion with graphs", *Numerische Mathematik*, vol. 1, p. 269–271, 1959.

[EVE 79] EVEN S., *Graph algorithms*, Comp. Sci. Press, 1979.

[FLO 62] FLOYD R., "Algorithm 97: shortest path", *Commun. ACM*, vol. 5, num. 6, p. 345, 1962.

[GAB 94] GABREL V., Méthodologie pour la planification de production de systèmes d'observation de la terre par satellites aspects algorithmiques et multicritères, PhD Thesis, LAMSADE, Université Paris-Dauphine, 1994.

[GAB 97] GABREL V., MOULET A., MURAT C., PASCHOS V., "A new single model and derived algorithms for the satellite shot planning problem using graph theory concepts", *Ann. Oper. Res.*, vol. 69, p. 115–134, 1997.

[GAR 79] GAREY M. R., JOHNSON D. S., *Computers and intractability. A guide to the theory of NP-completeness*, W. H. Freeman, San Francisco, 1979.

[GON 85] GONDRAN M., MINOUX M., *Graphes et algorithmes*, Eyrolles, 1985.

[HAI 85] HAIMOVITCH M., KAN A. H. G. R., "Bounds and heuristics for capacitated routing problems", *Math. Oper. Res.*, vol. 10, p. 527–542, 1985.

[HAL 93] HALLDÓRSSON M. M., "A still better performance guarantee for approximate graph coloring", *Inform. Process. Lett.*, vol. 45, num. 1, p. 19–23, 1993.

[HAR 69] HARARY F., *Graph theory*, Addison-Wesley, Reading, MA, 1969.

[HÅS 99] HÅSTAD J., "Clique is hard to approximate within $n^{1-\epsilon}$", *Acta Mathematica*, vol. 182, p. 105–142, 1999.

[JAI 85] JAILLET P., Probabilistic traveling salesman problem, Technical Report num. 185, Operations Research Center, MIT, Cambridge Mass., USA, 1985.

[JAI 88a] JAILLET P., "*A priori* solution of a traveling salesman problem in which a random subset of the customers are visited", *Oper. Res.*, vol. 36, num. 6, p. 929–936, 1988.

[JAI 88b] JAILLET P., ODONI A., "The probabilistic vehicle routing problem", in GOLDEN B. L., ASSAD A. A., Eds., *Vehicle routing: methods and studies*, North Holland, Amsterdam, 1988.

[JAI 92] JAILLET P., "Shortest path problems with node failures", *Networks*, vol. 22, p. 589–605, 1992.

[JAI 93] JAILLET P., "Analysis of probabilistic combinatorial optimization problems in euclidean spaces", *Math. Oper. Res.*, vol. 18, num. 1, p. 51–70, 1993.

[JAN 97] JANSEN K., "Approximation results for the optimum cost chromatic partition problem", in PARDALOS P. M., DU D., Eds., *Theoretical computer science, network design: connectivity and facilities location*, p. 143–168, 1997.

[JOH 74] JOHNSON D. S., "Approximation algorithms for combinatorial problems", *J. Comput. System Sci.*, vol. 9, p. 256–278, 1974.

[KAR 72] KARP R. M., "Reducibility among combinatorial problems", in MILLER R. E., THATCHER J. W., Eds., *Complexity of computer computations*, p. 85–103, Plenum Press, New York, 1972.

[KAR 98] KARGER D., MOTWANI R., SUDAN M., "Approximate graph coloring by semidefinite programming", *J. Assoc. Comput. Mach.*, vol. 45, num. 2, p. 246–265, 1998.

[KRU 56] KRUSKAL J. B., "On the shortest spanning subtree of a graph and the traveling salesman problem", *Proc. Amer. Math. Soc.*, vol. 7, num. 1, p. 48–50, 1956.

[LAD 75] LADNER R. E., "On the structure of polynomial time reducibility", *J. Assoc. Comput. Mach.*, vol. 22, p. 155–171, 1975.

[LAW 85] LAWLER E. L., LENSTRA J. K., RINNOOY KAN A. H. G., SHMOYS D. B., Eds., *The traveling salesman problem: a guided tour of combinatorial optimization*, Wiley, 1985.

[LEW 81] LEWIS H. R., PAPADIMITRIOU C. H., *Elements of the theory of computation*, Prentice-Hall, 1981.

[LOV 75] LOVÁSZ L., "Three short proofs in graph theory", *J. Combin. Theory Ser. B*, vol. 19, p. 269–271, 1975.

[MUR 97] MURAT C., Probabilistic combinatorial optimization graph-problems, PhD Thesis, LAMSADE, Université Paris-Dauphine, 1997.

[MUR 03] MURAT C., PASCHOS V., "The probabilistic minimum coloring problem", in BODLAENDER H. L., Ed., *Proc. 29th International Workshop on Graph Theoretical Concepts in Computer Science, WG'03*, vol. 2880 of *Lecture Notes in Computer Science*, Springer-Verlag, p. 346–357, 2003.

[NIC 99] NICOLOSO S., SARRAFZADEH M., SONG X., "On the sum coloring problem on interval graphs", *Algorithmica*, vol. 23, p. 109–126, 1999.

[PAP 81] PAPADIMITRIOU C. H., STEIGLITZ K., *Combinatorial optimization: algorithms and complexity*, Prentice Hall, New Jersey, 1981.

[PAP 94] PAPADIMITRIOU C. H., *Computational complexity*, Addison-Wesley, 1994.

[PAS 97] PASCHOS V., "A survey about how optimal solutions to some covering and packing problems can be approximated", *ACM Comput. Surveys*, vol. 29, num. 2, p. 171–209, 1997.

[PAS 03] PASCHOS V., "Polynomial approximation and graph coloring", *Computing*, vol. 70, p. 41–86, 2003.

[PAS 04] PASCHOS V., *Complexité et approximation polynomiale*, Hermès, Paris, 2004.

[PEE 99] PEETA S., ZHOU C., "Robustness of the off-line *a priori* stochastic dynamic traffic assignment solution for on-line operations", *Transportation Research Part C*, vol. 7, p. 281–303, 1999.

[SÉG 93] SÉGUIN R., Problèmes stochastiques de tournées de véhicules : un pas de plus vers le réalisme, Cahiers du CERO num. 3-4, CERO, Université de Montréal, 1993.

[SIM 90] SIMON H. U., "On approximate solutions for combinatorial optimization problems", *SIAM J. Disc. Math.*, vol. 3, num. 2, p. 294–310, 1990.

[STA 79] STATON W., "Some Ramsey-type numbers and the independence ratio", *Trans. Amer. Math. Soc.*, vol. 256, p. 353–370, 1979.

[TUR 41] TURÁN P., "On an extremal problem in graph theory (in Hungarian)", *Mat. Fiz. Lapok*, vol. 48, p. 436–452, 1941.

[VAZ 01] VAZIRANI V., *Approximation algorithms*, Springer, Berlin, 2001.

Index

I–L

M

N, O

P

R, S

T–W